The Study of Gene Action

Bruce Wallace and
Joseph O. Falkinham III

THE STUDY OF
GENE ACTION

Cornell University Press

ITHACA AND LONDON

First published 1997 by Cornell University Press.
First printing, Cornell Paperbacks, 1997.

Printed in the United States of America

Cornell University Press strives to utilize environmentally responsible suppliers and
materials to the fullest extent possible in the publishing of its books. Such materials
include vegetable-based, low-VOC inks and acid-free papers that are also either
recycled, totally chlorine-free, or partly composed of nonwood fibers.

Library of Congress Cataloging-in-Publication Data

Wallace, Bruce, 1920–
   The study of gene action / Bruce Wallace and Joseph O. Falkinham, III.
      p. cm.
   Includes index.
   ISBN 0-8014-3265-0 (cloth : alk. paper).—ISBN 0-8014-8340-9
(paper : alk. paper)
      1. Genetics—History.   2. Genes—History.   I. Falkinham, Joseph O.
II. Title.
QH428.W35   1997
576.5′09—dc21   96-52356

Cloth printing   10 9 8 7 6 5 4 3 2 1
Paperback printing   10 9 8 7 6 5 4 3 2 1

This book is dedicated to Will Durant:

The function of the professional teacher [is] clear. . . .
* to learn the specialist's language, as the specialist [has] learned nature's, . . .*
* to break down the barriers between knowledge and need, and*
* [to] find for new truths old terms that all literate people might understand.*
                                   —Will Durant, *The Story of Philosophy,* 1943

Contents

Preface   ix

1 Introduction   1

2 Inborn Errors of Metabolism   6

3 Research Organisms, Tools, and Procedures   12

4 Morphology   49

5 Color   76

6 Position Effect   104

7 Using the Environment as a Research Tool   131

8 Fate Maps: Studying Development through the Use of Mosaics   156

9 Transposable Elements   183

10 Tailoring Genes   212

11 Epilogue   241

References   249

Index   257

Preface

Genetics is a unique science. First, it has a beginning. In 1900, three plant breeders (Hugo De Vries, Karl Correns, and Erich Tschermak) independently determined the patterns of inheritance of certain characteristics of the garden pea, and each then acknowledged that these patterns had been observed and interpreted correctly thirty-five years earlier by an Austrian monk, Gregor Mendel. Second, genetics has an accepted research procedure, which can be inferred from my colleague M. M. Green's definition of a geneticist: "one who studies inheritance by making crosses." That is, once crosses between organisms become unnecessary for one's research on inheritance, one has become a physiologist, a biochemist, or a molecular biologist. One is no longer a geneticist.

Upon the rediscovery of Mendel's results in 1900, geneticists embarked on two avenues of research. The first was the identification of the gene. Mendel's "*Merkmalen*" (translated into English as "factors") presumably had a physical basis, as even Mendel recognized. But, where were they? And what were they composed of? Those questions are addressed (and answered) in a companion to this volume, *The Search for the Gene* (Wallace, 1992).

The second line of research, which continues today, investigates how genes carry out their activities. This avenue consists of the study of gene action, a study that has been severely limited by the prevailing technology of each successive era. The early study of gene action was, in fact, a study of external appearances that resulted from the *products* of gene action. Here, one must include the shape, color, and texture of organisms. Mendel's wrinkled and round peas, yellow and green peas, and tall and short pea plants are excellent examples. Some of these traits are not completely understood even now.

As biochemical techniques improved, experimental studies moved beyond external morphologies. Analyses of plant colors became possible, for example. Often such analyses were reduced to an understanding of pH and the colors assumed by indicator pigments. Enzymology was also involved in this era. One should not overlook here the classical (but once neglected) work of Archibald Garrod, who, in 1909, spoke of the inborn errors of metabolism.

With the discovery of DNA structure by James D. Watson and Francis H. C. Crick in 1953, our understanding of gene action and gene structure began to converge. Genetics at that moment was reduced to chemistry. How the gene was built determined to a great extent how it acted. Genetics remains an independent science only to the extent that systematic crosses are still required at times to obtain essential information. Otherwise, research proceeds by fractionating cells, by providing cellular homogenates with precursors (often labeled with radioactive isotopes), and by studying the kinetics of various reactions. In brief, the study of gene action has spilled far beyond the science of genetics—thanks, of course, to the excellence of many early geneticists.

In the pages that follow, the study of gene action is traced and presented as an intellectual exercise. The reader should remember that, however crude past studies now appear, early workers were not stupid. Rather, they lacked adequate technologies. Anyone who seeks to answer a question that lies beyond his or her technical skills appears, in hindsight, to have blundered. The seeming blunders in many instances should be viewed as deserving of commendation rather than as signs of incompetence. Even now, a large proportion of all publications contribute little or nothing to the solution of those problems to which they are ostensibly addressed. Perhaps that constitutes the strongest argument for exercising prudence in hurrying from the research laboratory to the hospital clinic—or to one's public relations office.

We wish to thank Meg Nugent for preparing the artwork and Sue Rasmussen for the exhausting task of typing not only the final manuscript but each of the preliminary versions, only to see them revamped. Helene Maddux, of Cornell University Press, and Margo Quinto have improved the manuscript considerably; we thank them for their effort as well.

BRUCE WALLACE

JOSEPH O. FALKINHAM III

Blacksburg, Virginia

1   Introduction

In his appendix to C. H. Waddington's (1957, p. 192) *The Strategy of the Genes,* Henry Kacser wrote,

> The great achievement of genetics has been the demonstration that certain portions of the chromosome control to a large extent the development and behaviour of organisms. Although we believe the elaborate circumstantial evidence that the genes act in this fashion, it is desirable that, like Justice, they should manifestly appear to act. This activity is primarily revealed in the single chemical and physical steps, which are the "atoms" of the organism. To demonstrate the genes in action, then, requires the language of the molecular calculus instead of the Mendelian arithmetic which revealed their presence.

Genetics involves two quasi-independent topics: the identification of the gene (i.e., of Mendel's *Merkmalen*) and the determination of the means by which genes accomplish their tasks. *The Search for the Gene* (Wallace, 1992) is devoted to the first topic, almost to the exclusion of the second. What a gene does, of course, reveals its presence much as odors, sounds, and slight disturbances of its surroundings reveal the presence of a field mouse to an alert cat. To that extent, the search for the gene touched on gene action; that is, only when that action revealed something concerning the nature of the gene.

If we ignore cul-de-sacs (of which there were many), we can describe the search for the gene as an ever-narrowing, essentially linear search through time. Many persons were involved; their lives overlapped; their studies were not always coordinated; the research

1

organisms of choice changed through time. Nonetheless, starting with the rediscovery of Mendel's paper and continuing well into the study of bacteriophage and plant viruses, geneticists relentlessly narrowed the possible sites for and the chemical structure of the gene. Their search terminated rather abruptly when James D. Watson and Francis H. C. Crick published their insights regarding the structure of deoxyribonucleic acid (DNA).

The study of gene action, unlike the search for the gene, cannot be fitted to a linear model. Rather, it can be viewed as a harp (Figure 1-1).

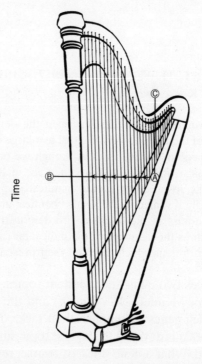

Figure 1-1. The harp as an analogy for the study of gene action. Unlike the search for the gene, which was essentially linear through time, the study of gene action is (at least) two-dimensional. The vertical direction represents time. The upward-sloping base of the harp (a vector) represents intellectual, procedural, and technological advances (bases of individual strings) through time. An advance, A, that is made at a given moment can be brought to bear (line AB) on preexisting problems (each longer string to the left). It also gives rise to questions and concepts that did not exist earlier (line AC) but that, in turn, become the subjects of investigation using still later techniques. Many of the comings and goings of research activity on questions of long standing are the consequences of newly discovered tools and techniques.

New techniques occur sequentially along the upward-sloping basal soundboard. Each gives rise to a vertical string that represents the questions that are now possible to ask and the concepts that are now feasible to entertain. At the same time, however, new techniques can be applied to preexisting problems (the longer strings), including the very earliest ones raised by Gregor Mendel's studies. For example, on September 22, 1915, Calvin B. Bridges discovered among his flies (*Drosophila melanogaster*) a single individual whose halteres exhibited a winglike (rather than a clublike) appearance. The abnormal phenotype proved to be caused by a recessive mutant gene located at map position 58.8 of chromosome 3. Because Bridges was unfamiliar with the embryonic development of *Drosophila* and thought (erroneously) that his abnormal fly had grown a second thorax, he designated this mutant *bithorax*. During the years immediately following its discovery, *bithorax* was used only as a tool in extending knowledge concerning linkage and other aspects of classical genetics. Facets of genetics other than Mendelian ratios and linkage data became available for study as more *bithorax*-like mutants were discovered: other *bithorax* alleles in 1925 and 1934; *bithoraxoid* (1919), *postbithorax* (ca. 1954); *Contrabithorax* (1949); and *Ultrabithorax* (1934). With these many alleles, not only could a fine-structure analysis of this chromosomal region be carried out (Lewis, 1951), but also some grasp of the role of chromosomal pairing in affecting gene action could be investigated (Lewis, 1955). The latter study, of necessity, awaited the discovery (Painter, 1931) and detailed band-by-band mapping (Bridges, 1935) of the giant chromosomes of larval salivary gland cells.

The need to await the development of ever finer or more powerful investigative techniques is a characteristic of the study of gene action. Progress may be rapid or slow; seldom does science advance at a uniform pace. Conceptual advances, in many cases, awaited technical ones, but they did not always follow immediately. The use of garden peas in the study of heredity was essentially a technical advance promoted by Mendel; nevertheless, thirty-five years were to elapse before his colleagues and successors grasped what he had so patiently explained. In science, a burst of frenzied activity can, at times, conceal a mere spinning of wheels or running in place. Following the rediscovery of Mendel's publication in 1900, most genetic research consisted of confirming and reconfirming Mendel's numerical results. Today, the ability to determine the base sequence of DNA fragments provides the molecular geneticists with a remarkable tool, one that has changed the character of genetics. Much DNA sequencing (e.g., the sequencing of the entire genomes of humans, mice, flies, nematodes, many plants, and microorganisms) is being carried out because (1) the technique for doing so exists and (2) both the rules and the tools for interpreting those sequences are now available.

In recent years, the study of the *bithorax* locus has moved into the realm of molecular biology and genetic engineering. Electron microscopy has revealed (as the light microscope could not) that some morphological abnormalities involve the orientation of developmental axes: dorsal-ventral as well as anterior-posterior. The ancient mutant *bithorax* clearly remains the object of considerable modern research.

One might note in passing that the basis for the wrinkling of Mendel's wrinkled peas is now known: The gene that specifies an enzyme that, in turn, generates branched starch molecules malfunctions in wrinkled peas. The failure to produce branched starch molecules has consequences bearing on starch, lipid, and protein synthesis in the seed—consequences that lead to wrinkled rather than smoothly spherical peas.

Experimental techniques are discovered and improved upon in linear, temporal sequence. A new technique can be utilized in the study of an old problem only after it has been developed and perfected. This obvious fact determines the format of the chapters that follow. Beginning with Chapter 4, we attempt to discuss research on the action of genes in a sequence approximately corresponding to that with which investigative procedures (often crude) were applied: morphological examinations, the production of phenocopies, the utilization of position effect, and the like. The other axis forming the intellectual landscape is chronological time, starting with 1900 because in that year Mendel's paper was rediscovered. The individual chapters, consequently, traverse time. The early chapters cover the greatest lengths of time, with periods of great activity alternating with slower periods during which workers awaited the discovery of a new experimental approach; the intellectual contributions of old approaches become exhausted rather quickly. The comings and goings of research activity in each chapter, then, resemble the swellings in numbers and the subsequent dwindlings of various organisms in the geologic record.

Chapter 3 is devoted to a review of techniques and the times of their discovery or invention. Phenocopies were a favorite tool of Richard Goldschmidt and Walter Landauer during the 1930s and 1940s. Studies of the artificial induction of mutations awaited H. J. Muller's and L. J. Stadler's observations on X rays during 1927 and Charlotte Auerbach's work on mustard gas during World War II. Experiments (other than exploratory ones) based on the structure of DNA, of necessity, follow James D. Watson and Francis H. C. Crick's publications of 1953. The rapid sequencing of DNA fragments did not become possible until 1975. The identification of seminal advances in techniques and the positioning of these advances in time will be the tasks that are undertaken in Chapter 3.

Moving backward, crablike, we come to Chapter 2, a chapter devoted entirely to one person: Archibald E. Garrod (1857–1936). With respect to gene action, Garrod occupies a position analogous to that occupied by Gregor Mendel with respect to the search for the gene itself. Garrod, even before Mendel's paper was rediscovered, was studying the pigmentation of human urine, as well as familial expression of other traits such as rheumatoid arthritis. An accomplished biochemist, Garrod had learned much about black urine and had written about the individuality of persons and the impact that such individuality has on the practice of medicine. There is, in our minds, no means by which Garrod could have been woven into the general outline of this introductory chapter in a manner that does him justice. Consequently, Chapter 2 is devoted to him and the Croonian lectures, delivered before the Royal College of Physicians of London in June 1908, that were the basis of his book *Inborn Errors of Metabolism* (Garrod, 1909).

2 Inborn Errors of Metabolism

During the early 1940s, George W. Beadle, Edward L. Tatum, and their colleagues intensively studied the bread mold *Neurospora crassa* with respect to the biochemical requirements of radiation-induced mutations. In brief, they discovered (as they expected they would) that derivatives of irradiated wild strains of this mold were often unable to grow on minimal medium, a medium that supplies the bare necessities required by wild-type cultures: a carbon source; a nitrogen source; inorganic salts that provide sulfur, potassium, calcium, and phosphorus; and the vitamin biotin. Stepwise enrichment of the insufficient minimal medium through the addition of vitamins or amino acids, for example, revealed that a mutant strain, at times, could be cultured successfully. Subsequent tests involving the addition of single amino acids or single vitamins were then able to pinpoint the mutant lesion. These genetic techniques led to the "one gene–one enzyme" hypothesis: namely, the notion that the normal allele at each mutant locus is responsible for specifying one, and only one, enzyme. Because it lacks this enzyme, the mutant strain requires the addition of the enzyme's normal end product for growth. The ingenious combination of genetic and biochemical analyses provided a powerful means for revealing the numerous individual steps in complex metabolic pathways.

In a review article titled "Biochemical Genetics," Beadle (1945, p. 25) said:

From a historical standpoint it is a curious fact that until recent years alkaptonuria has played almost no part in the development of theories of gene action. Garrod's book . . . and its second edition . . . treated alkaptonuria in detail from both the biochemical and the genetic points of view. This book . . . should be credited as representing the beginning of biochemical genetics, but unfortunately, until its significance was pointed out recently, particularly by Haldane, it has remained practically unknown to geneticists. The biochemists, it should be said, were less guilty of neglect, but they apparently were not prepared to appreciate fully the genetic implications.

This statement reveals the enormous divide that separated early geneticists and biochemists. The latter saw the former as persons who were fascinated with numerical ratios divorced from material substances; those geneticists who inferred from their data the necessary characteristics of any physical substance that might underlie inheritance patterns spoke a language that was incomprehensible to biochemists. Small wonder, then, that H. J. Muller (1929) could complain that "physiologists" lacked both the tools and the desire to unravel the mysteries surrounding the material basis of inheritance.

In his Croonian lectures, Garrod (1909) touched on both genetics and biochemistry. In the study of alkaptonuria ("black urine") for example, he noted (p. 24): "Of the cases of alkaptonuria . . . a very large proportion have been in children of first cousin marriages." Upon tabulating cases of the disorder known to him, he determined that of 18 families studied, 8 (44%) were the children of parents who were first cousins. The remaining 10 were families of parents who were not related. The proportion of cousin marriages among the parents of alkaptonurics was "altogether abnormal"—that is, much greater than the (approximately) 3% incidence of cousin marriages that was characteristic of Great Britain at that time.

Having dismissed the notion that anomalies may arise de novo merely because parents are of one blood, Garrod (1909, p. 25) observed that "it is obvious that the reappearance of a latent character which both parents tend to transmit is likely to be favoured by the mating of members of certain families." He was already prepared to regard alkaptonuria as a "rare recessive character in the Mendelian sense," as had been earlier proposed by both William Bateson and R. C. Punnett (Garrod, 1909, p. 26).

This venture into population genetics by Garrod can be elaborated upon even though to do so briefly interrupts our narrative: If alkaptonuria is caused by a rare recessive allele whose frequency in a population is q, the frequency of homozygous alkaptonurics among children of nonrelated parents is q^2. Because the children of cousin marriages are homozygous (through inbreeding) for $1/16$ of their entire genome, the frequency of alkaptonurics among children of cousin marriages is

$$\frac{q}{16} + \frac{15q^2}{16},$$

or

$$\frac{q}{16}(1 + 15q).$$

Following Curt Stern's (1960) account, let c equal the frequency of cousin marriages among all marriages; then the frequency of alkaptonurics from cousin marriages equals

$$\frac{cq}{16}(1 + 15q).$$

The proportion, k, of alkaptonuric children from cousin marriages among all alkaptonuric children equals

$$\frac{\frac{cq}{16}(1 + 15q)}{q^2},$$

or

$$k = \frac{c}{16}\left(\frac{1}{q} + 15\right).$$

Clearly, the smaller the value of q, the greater must be the proportion of cousin marriages among the parents of afflicted children. Garrod's inference was exactly on target.

In attempting to compare the proportion of alkaptonurics with that of normal sibs on the basis of data collected from 18 families, Garrod neglected to make a simple but necessary correction. Among his families there were 65 normal sibs and 34 who exhibited alkaptonuria. As he admitted (p. 28), the relative numbers of dominant and recessive offspring "departed widely from those required by Mendel's law." Garrod (1909) then proceeded to enumerate possible sources of error. Not noted by him, however, was the need to remove from his data the 18 alkaptonuric children (*propositi*) who brought affected families to his attention; Mendel's law (the expected 3:1 ratio of normal to alkaptonuric children) applies only to the sibs of these *propositi*. After adjusting Garrod's numbers to 65 normal and 16 alkaptonurics among 81 children, one sees that the expected numbers are approximately 60 normals to 20 alkaptonurics. The observed numbers now fit Mendelian expectations extremely well.

Figure 2-1 outlines the metabolism of phenylalanine and tyrosine in humans; it is adapted from a well-known textbook of the 1950s, *General Genetics* (Srb and Owen, 1952). The credit line beneath the Srb and Owen figure cites Beadle (1945). Beadle (1945) cites J. B. S. Haldane (1942, p. 47), who referred to "the remarkable work of Garrod." Many geneticists of the 1950s, having failed to follow that trail, unthinkingly assumed that the "pioneering" work done by Beadle and Tatum on the biochemical genetics of *Neurospora* was the first to elucidate the metabolism of phenylalanine and tyrosine despite Srb and Owens's reference to Garrod. To undermine that mistaken belief, we have inserted the page numbers of Garrod's (1909) first edition of *Inborn Errors of Metabolism* into both the chemicals themselves (indicating that Garrod knew the chemical formulae) and into the transitional pathways (indicating that Garrod understood the chemical transformations, as well). The date, 1891, that is appended to the formula of homogentisic acid reveals that it was "isolated, analyzed, and fully investigated by Wolkow and Baumann" by that year (Garrod, 1909, p. 48).

The knowledge gained by Garrod concerning the biochemical aspects of phenylalanine and tyrosine metabolism, including the excretion of homogentisic acid by alkaptonurics, was based largely on feeding experiments. Feeding tyrosine to alkaptonurics greatly increased the output of homogentisic acid. A corresponding increase followed an augmented intake of protein, especially of proteins rich in aromatic amino acids. Because tyrosine did not appear to account for all the homogentisic acid excreted, phenylalanine was also impli-

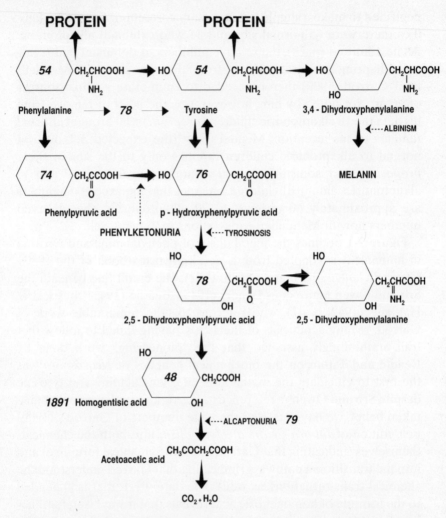

Figure 2-1. The scheme for the metabolism of phenylalanine and tyrosine that appeared in Srb and Owen's (1952) *General Genetics;* these authors cited Beadle (1945) as their source. The numbers that appear in the benzene rings of six compounds refer to the pages in Garrod's *Inborn Errors of Metabolism* (1909) on which he cites these same structures; other page numbers reveal that he was also aware of the metabolic pathways within this scheme. The formula for homogentisic acid was, according to Garrod, known to Wolkow and Baumann by 1891.

cated. Feeding experiments confirmed that phenylalanine did indeed serve as a parent substance. The increased excretion of homogentisic acid virtually equaled in quantity the amount of phenylalanine and tyrosine fed to an alkaptonuric. Feeding tryptophan, also an aromatic amino acid, was shown not to influence the excretion of homogentisic acid.

The many details of Garrod's research need not be recounted here. Suffice it to say that a bacterial origin of homogentisic acid was ruled out. Further, he demonstrated that homogentisic acid is not an abnormal metabolic product but, rather, a normal one that, in alkaptonurics, escapes destruction. By taking a sufficient quantity (8 g) of homogentisic acid, a normal person can induce a temporary alkaptonuria; lesser amounts are metabolized and not excreted. Homogentisic acid perfused through liver tissue yields acetone. Finally, Garrod proposed the following (p. 73):

> [If] homogentisic acid is the product of normal metabolism, the result of the administration of various aromatic acids to alkaptonurics may reasonably be expected to throw light upon the higher intermediate steps between the parent protein fractions and that substance. Any compound, which represents a link in the chain should, on the one hand, be destroyed, as tyrosine and homogentisic acid are, in the normal organism, and, on the other hand, should increase the output of the latter by alkaptonurics. Any substance which does not behave in the manner indicated cannot form such an intermediate link.

The pathways shown in Figure 2-1 are the outcomes of numerous painstaking feeding experiments; not shown in the figure are the numerous substances that were tested and found not to behave "in the manner indicated."

Just as geneticists since 1900 have paid homage to Gregor Mendel as the father of genetics, so have we paid homage here to Archibald Garrod for the insights that he showed in his studies of alkaptonuria, albinism, cystinuria, and pentosuria—a few of the inborn errors of human metabolism with which he was concerned. He even assessed properly the lack of melanin in albinos (p. 39): "What the albino lacks is the power of forming melanin, which is normally possessed by certain specialized cells. . . . An intracellular enzyme is probably wanting."

3   Research Organisms, Tools, and Procedures

The following question was posed in the "Amateur Scientist" column of a now all-but-forgotten issue of *Scientific American:* How many games must be scheduled for an elimination tournament involving 64 ball teams? The author went on to say that the correct solution can be obtained by summing 32 (1st round), 16 (2nd round), 8 (3rd round), 4 (4th round), 2 (semi-finals), and 1 (final game); the total equals 63 games. The solution has been arrived at by sheer effort. A much neater solution is obtained by noting that every team but one must lose once, and only once; hence, 63 games are required. Such intellectual gamesmanship is often presented as the contrast between the reasoning of mathematicians and physicists. The moral can be extended, however: success in science frequently depends upon the finesse with which research is carried out; *a need for sheer labor can thwart scientific inquiry.* Snobbery, of course, plays its role. Once, at a symposium the third speaker (who shall not be identified here) declared that he would "now show that the results of the previous two speakers could have been foretold by logic alone." There is no substitute for intellect and insight, to be sure; however, there are also no good substitutes for the proper experimental organism and efficient experimental procedures, as well.

Science as an enterprise depends upon puzzling observations that pique someone's curiosity and the existence of procedures that allow one to distinguish between alternative possible explanations (hypotheses). Without an observer's sense of puzzlement, there would be no science. There would be no need for explanations. The

Figure 3-1. The varied patterns of different species of sparrows. Each pattern is determined by the bird's genetic program. An observer may legitimately ask, How are the genes that are responsible for the pigmentation of feathers turned off and on in such a precise manner? What is the purpose of such diversity in pigmentation? is a second, equally legitimate question. The latter is a question that touches on behavior, communication, and reproductive isolation. The former is a question that concerns gene action. Pattern variation within a species (melanic variants, for example) raises questions concerning transmission genetics, a subject that concerned Gregor Mendel.

diagrams in Figure 3-1 illustrate the pigmentation patterns on the heads of several small birds (sparrows). These patterns must evoke questions (Why? How?) if they are to lead to scientific investigations. To accept the patterns as given removes them from the realm of science.

Of necessity, the study of gene action awaited a convincing demonstration that inheritance patterns are reproducible from generation to generation and, therefore, must have a physical explanation. Mendel is generally regarded as the one who provided that demonstration. The next step, identifying the responsible physical elements, posed one problem. Determining how they perform their functions (and how to determine how they do so) posed a second. The success or failure of the investigations into these matters rested largely on the choice of a research organism. The organism and the manipulations to which it could be subjected opened some areas of research, while closing others.

The physical tools that are available to the research scientist determine in large measure what manipulations can be performed and what observations can be made. The status of ancillary fields—physics, chemistry, and engineering—has greatly influenced genetic (and other biological) research; automation has become especially important in carrying out repetitive procedures. Still, it is the individual investigator who detects weaknesses in conventional wisdom, who poses unconventional questions, and who designs the experiments that bear on these questions.

Errors in experimental design or oversights in logic lead to erroneous conclusions. Garrod's failure to adjust the numbers of normal and alkaptonuric sibs by removing the 18 alkaptonuric *propositi* led him *not* to confirm Mendel's law. The path of science is littered with wrongly designed experiments. The surface tension of the evaporating mounting fluid overturned vertical-standing bacteriophage so that they "faced" the much larger bacteria in electron micrographs, thus creating a short-lived belief that, like tadpoles, individual phage particles were attracted by, and "swam" toward, the bacterium over a considerable distance. That error was corrected when liquid carbon dioxide was used as the mounting fluid; at a particular combination of pressure and temperature, carbon dioxide passes from a gaseous to a liquid state with no intervening "surface." A. D. Hershey, one of the original members of the Phage Group at Cold Spring Harbor, once received a card from a French virologist acknowledging that he (Hershey) had asked and answered questions that the writer had never thought to ask! That was high praise, indeed!

Research organisms

The choice of an experimental organism rests on many considerations, the chief of which may be ease of analysis. Mendel (1865) listed his reasons for choosing peas: Though peas are normally self-pollinating, emasculation and cross-pollination make it easy to perform hybridizations. Many true-breeding varieties were available for study. With respect to characters of the cotyledons (yellow versus green; round versus wrinkled), peas in the pod represent the "next" generation; therefore, many more individuals can be scored for these traits than for those that are apparent only in mature plants (e.g., height). Because a quasi-historical approach has been taken in this book, research organisms are presented approximately in their order of appearance through time.

Drosophila

Most elementary textbooks enumerate the advantages possessed by the vinegar fly, *Drosophila melanogaster:* ease of culturing, short generation time, ease in collecting virgin parents and in making single-pair matings, and the production of many (100 or more) progeny by individual females. These are the immediately attractive characteristics of *D. melanogaster*—a physically strong and robust species of fly. Workers using other *Drosophila* species have often found to their dismay that their flies are delicate, easily injured or killed, require special culture conditions, and are not highly productive. In this paragraph only the characteristics of *D. melanogaster* that attracted early workers have been described; later, we will note, as research procedures improved, how even the complexity of insect development became amenable for study in these flies. Although laboratory strains of *D. melanogaster* had been maintained at Harvard University by W. E. Castle since 1903, the flood of papers dealing with the genetics of these flies that emanated from T. H. Morgan's group at Columbia University began in 1910.

Mammals

Many early geneticists studied laboratory mice, rats, rabbits, and guinea pigs. These persons concentrated primarily on problems of interest to medicine and mammalian physiology. Publications dealing with the formal genetics of laboratory mammals first appeared about 1903. The strains of inbred mice maintained by the Jackson Laboratory, Bar Harbor, Maine, provide much of the material needed for research carried out at and under the financial support of the National Institutes of Health. Sewall Wright, one of the founders of population genetics, worked with guinea pigs for many years (1915–1925), first as a geneticist with the U.S. Department of Agriculture and later at the University of Chicago. Careful attention to mutant coat colors, and especially of the pigmentation of animals carrying different combinations of mutant genes, allowed Wright to speculate on pigment formation—a matter that otherwise lies in the realm of biochemistry.

Plants

Plants of exceptional economic value were the chosen research organisms of many investigators. Indian corn (*Zea mays*) has been studied by geneticists at virtually every land grant university. These many studies have dealt not only with the quantitative genetics of corn (hybrid vigor in this species was described by Charles Darwin [1876, table 97] and studied extensively at Cold Spring Harbor by G. H. Shull [1908]) but also with its "formal" genetics. An early center for the latter studies was Cornell University, where R. A. Emerson acquired an outstanding group of students during the 1920s. One of these students, Barbara McClintock, was to receive the Nobel Prize in 1983 for her work on genetic elements that control the expression of other genes and that move about within the corn plant's genome. Such genetic elements are now the subject of intense molecular research not only in corn but also in other organisms, including microorganisms and human beings, in which similar phenomena have been recognized.

Higher plants are complex organisms but, once one understands their life histories, they offer an investigator many advantages: Their cells, once formed, remain in place, thus easing the study of developmental problems, especially patterns of growth (see Figure 3-2). Individual leaves can often provide ample material for biochemical and physiological analyses. Pollen grains represent haploid progeny of the parental plant; large numbers of these can be examined for genetic differences (Figure 3-3). Pollen may also be germinated in culture, thus producing haploid plants. Whole plants of many plant species can be regenerated from single protoplasts, as well. The interactions among nuclear, chloroplast, and mitochondrial genes provide the plant geneticist with problems that are not encountered within animal systems, whose cells lack chloroplasts.

At the present time, a small plant (*Arabidopsis thaliana*) has emerged as the *Drosophila* of the plant world. It has a short generation time (about 50 days), can be cultured on synthetic medium in test tubes, is easily hand pollinated, and produces as many as 10,000 seeds per plant. Furthermore, it is a true diploid (many plants carry

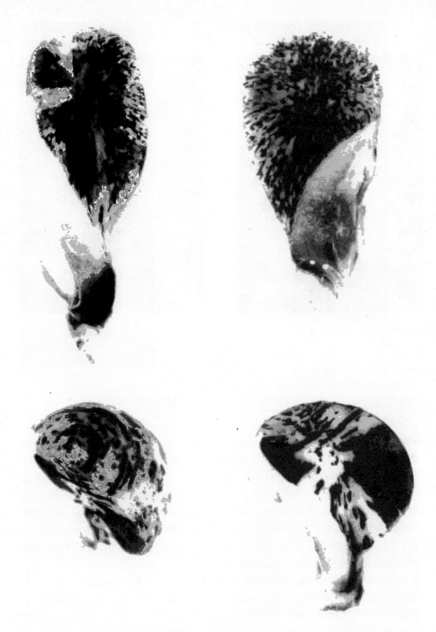

Figure 3-2. Variation in the developmental patterns of different corn kernels as revealed by clones of cells that possess (for genetic reasons that need not be explained here) starch granules that stain differently when exposed to an iodine solution. Notice that, in the upper kernels, the darkly pigmented areas lead in well-defined rays from an initial focal area. In the lower kernels, the darkly stained clones of cells reveal, instead, what appears to be a nested series of cones. (Redrawn from McClintock, 1978, by permission of Academic Press.)

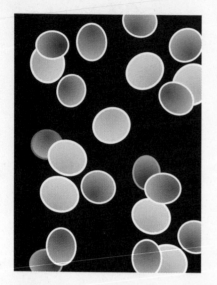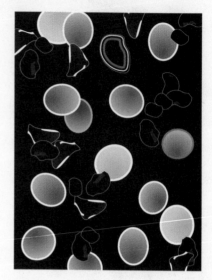

Figure 3-3. Normal and aborted pollen of *Zea mays*. On the left are pollen grains of a normal plant; on the right, the normal and aborted pollen of a plant that was heterozygous for a reciprocal chromosomal translocation. The starch in pollen may, for genetic reasons, stain either red or blue with an iodine solution; thus, large numbers of iodine-stained pollen may be screened visually for starch mutants.

genomes and genome fragments of numerous ancestral species) with an extremely small genome size; single genes perform functions in all tissues of *Arabidopsis* that in other higher plants are carried out in different tissues by different, segregating genes that are scattered throughout their genomes. An important feature for geneticists, one not possessed by *Drosophila* but one that is characteristic of *Neurospora* and yeast, is the opportunity provided by the organism for recovering all four products of meiosis occurring in a single cell. A mutant strain of *Arabidopsis* provides this opportunity (Figure 3-4), because the four pollen grains arising from each ancestral spermatocyte remain glued together.

Neurospora

In 1927, two mycologists, C. L. Shear and B. O. Dodge, described the life history and mating behavior of *Neurospora crassa* (which they called *Monilia sitophila*), the red bread mold. Because of his under-

Figure 3-4. Two clusters of four pollen grains each that characterize a mutant (*quartet*) strain of *Arabidopsis thaliana*. Each cluster provides the geneticist with all of the genetic material resulting from meiosis within a single spermatocyte. (Reprinted with permission from D. Preuss, S. Y. Rhee, and R. W. Davis, "Tetrad Analysis Possible in *Arabidopsis* with Mutation of *QUARTET* (*QRT*) Genes," *Science* 264 [1994]: 1458–1460. Copyright 1994 American Association for the Advancement of Science.)

standing of genetics, Dodge unraveled the mating-type system and demonstrated the segregation of Mendelian characters among ascospores. He then brought *Neurospora* to the attention of T. H. Morgan; it was in the latter's Cal Tech laboratory that many of the problems associated with the formal genetics of *Neurospora* were solved. Beadle and Tatum, whom we have mentioned earlier, became interested in *Neurospora;* J. R. S. Fincham (1985, p. 4) wrote: "The attraction of *Neurospora,* with its excellent genetics and rapid life cycle, was enhanced by the demonstration that it had no requirements for organic nutrients other than biotin and sugar (or other carbon source)." *No requirements for organic nutrients!* Stated differently (and in a manner seldom emphasized by most biologists of that era), every complex organic molecule that occurs in *Neurospora* is made by the mold itself. The ability to do so, in the minds of Beadle and Tatum, lay in the mold's genetic material. Beadle and Tatum had at their disposal an organism that, when carrying a genetic lesion produced by either UV- or X-radiation as a mutagenic agent, might fail to grow on minimal medium. Such mutant strains often grew, however, on media to which certain chemical substances were added. Logic suggested (recall Garrod's earlier but largely unrecognized work) that these substances were those lying in the metabolic pathway subsequent to the point blocked by the mutant gene. Indeed, the early work on mutant *Neurospora* was of great interest to biochemists, for it provided them with a new approach for deducing the sequence of reactions in metabolic pathways. These same biochemists were often less interested in understanding either the structure of or the mode of action of the gene itself.

Bacteria

Bacteria such as *Escherichia coli,* a common colon bacillus, also possess the ability to synthesize all needed organic compounds from simple, even inorganic, precursors. Many geneticists undertook bacterial studies; the attraction coincided with but was not necessarily caused by World War II. The chief attraction, especially to the mathematically literate biologists, lay in the *individuality* of bacteria; they do not come as complex organisms as do higher plants and animals or even as a multinucleate syncytium as does *Neurospora.* Enormous numbers of individuals can be easily grown, counted, and studied. Rare mutants that occur only once in a million individuals can be recovered by the hundreds in culture tubes that may contain as many as 10^8–10^9 (100 million to 1 billion) bacteria.

Two laboratories—those of Beadle and Tatum of Cal Tech and Salvadore Luria and Max Delbrück (associated with Cold Spring Harbor)—initiated genetic studies using *E. coli.* Beadle and Tatum were interested primarily in elucidating biochemical pathways by means of mutant (*auxotrophic*) strains; Luria and Delbrück were interested in demonstrating that bacterial mutations were equivalent to the mutations geneticists had studied and were studying in higher organisms. The now-classic paper by Luria and Delbrück (1943) demonstrated that mutations conferring resistance to bacteriophage in *E. coli* did not occur *in response to* a given challenge but, rather, preexisted in the bacterial population. Bacteriophage, in this study, merely provided the means by which Darwinian (rather than Lamarckian) selection was confirmed. The two early branches of *E. coli* research converged in Cold Spring Harbor, where Milislav Demerec (director of both the Department of Genetics and the Biological Laboratory) and his many younger colleagues carried out intensive genetic studies on *E. coli* and its near-relative *Salmonella typhimurium*. The convergence was perhaps consummated when Joshua Lederberg and E. L. Tatum succeeded in obtaining genetically recombinant individuals from mixed cultures of two multi-mutant bacterial strains (Lederberg and Tatum, 1946). The Demerec research group were the first to reveal in *E. coli* the remarkable correlation between sequences of biochemical steps in a metabolic pathway and the arrangement of the genes controlling those steps in the bacterium's linkage map (Demerec and Hartman, 1959); this correlation (later found in other bacteria, as well) became intelligible as the role of DNA and RNA in protein synthesis became clear.

Bacteriophage

Bacteriophage were already a genetic tool by 1943. Under Luria and Delbrück's influence, a school of phage workers formed at Cold Spring Harbor—a school soon joined by A. D. Hershey and supplemented by disillusioned nuclear physicists and by bright graduate students (of whom James Dewey Watson was one). The "phage course" at Cold Spring Harbor had as its first students well-established workers who returned to their university laboratories and transmitted their newly acquired skills to their students.

The choice of bacteriophage for genetic studies was based on essentially the same considerations that had favored *Drosophila melanogaster* a half-century earlier: ease of culture and short (20 minutes) generation time. In addition, single-particle infections could be made with ease, and each infected bacterium produced numerous (100 or more) progeny phage. The ability of the investigator to handle both the host bacteria and the bacteriophage as collections of *indi-*

viduals, adjusting the concentration (i.e., the number) of each to suit the needs of the particular experiment and exposing either or both to prior chemical or other experimental treatment (including mutagenic radiation), led to the solution of problems that other workers were unable to pose. For these reasons, bacteriophage workers outstripped those working with plant viruses despite the crystallization of the tobacco mosaic virus by Wendell Stanley in 1933 (see Stanley, 1935).

Because research on bacteriophage appeared to have completed its contribution to the study of genes and gene action, many pioneer phage workers deserted this organism in search of new and more challenging research problems: Delbrück, phototaxis in the fungus *Phycomyces;* Renato Dulbecco, mammalian tissue culture and animal viruses; Crick, the nervous system; Seymour Benzer, development in *Drosophila;* and Sydney Brenner, nervous system development in the nematode *Caenorhabditis elegans.* Many now study developmental (homeotic) mutants in *Drosophila.* This organism has entered a true Renaissance because modern molecular techniques provide answers to questions that earlier workers may have asked but, by being premature, were unable to answer.

Yeast

Studies of Baker's yeast (*Saccharomyces cerevisiae*) grew out of the investigations of *Neurospora,* stimulated in large measure by yeast's commercial value in brewing, wine production, and baking. Although lacking the ordered tetrads of *Neurospora,* they grow as, and can be manipulated as, individual cells. (As an aside, one might note that an Anheuser-Busch employee [William S. Gossett] studying yeast cells by means of a hemocytometer derived the Poisson distribution that describes the expected distribution of yeast cells within the "checkerboard" background of the instrument; this—and other—analytical procedures were published by him under the pseudonym "Student.")

A French geneticist, Boris Ephrussi (1952), described small-colony-forming (*petite*) variants that proved to be incapable of respiration; this work contributed to the discovery of mitochondrial (i.e., nonnuclear) DNA. Studies on yeast have also led to an understanding of properties that maintain the integrity of chromosomes—their replication, segregation, and resistance to degradation. This knowledge has led in turn to the ability to create artificial chromosomes, an ability of considerable importance in the study of large DNA fragments.

Nematodes

Special mention must be made here of Sydney Brenner and his deliberate search (begun in 1963) for an extremely simple higher organism that could be cultured on a petri dish—a metazoan to which one could apply many bacteriological culturing and plating methods. He found his organism: the nematode *Caenorhabditis elegans.* This small worm has a constant number (<1000) of somatic cells, about one-third of which (also a constant number) constitute the nervous system. Today, the developmental origin of every cell is known, as is the time (following fertilization) of every cell division. The neuronal connections have been mapped by the use of electron micrographs. Other cell-cell connections are being disclosed by following the movement of intracellular dyes. Work has begun on sequencing the worm's entire genome, which contains as much DNA as a single average-sized human chromosome. This impressive progress has occurred within the brief span of a quarter of a century.

Humans

Early in the nineteenth century, Alexander Pope declared that "the proper study of mankind is man." And, indeed, one of the first evidences of heredity (ca. 1740) was the occurrence of extra fingers on one or more members of one German family in each of four successive generations. Given the infrequency of this trait among the local inhabitants, Pierre Maupertuis calculated the odds that this one family should be so afflicted by chance alone: 1 in 8×10^{12}! Thus, Maupertuis argued, this family possessed a germinal flaw that was transmitted from parent to child, generation after generation.

Familial analyses performed since the early 1900s have revealed many human traits that appear to be transmitted as Mendelian genes. Unfortunately, family sizes are small, and, as a result, many similar-appearing families must often be combined. In addition to possible statistical errors (see p. 9), errors in genetic analysis may result. Mutant genes with similar phenotypic effects can be incorrectly regarded as due to lesions at a single locus (e.g., hemophilia is now seen as a phenotype resulting from single mutations at many gene loci). Other traits (e.g., wanderlust, love of the sea, and leadership), have been ascribed to genes for little or no reason. Simplistic and slip-shod studies of the latter type led to many of the unfortunate social programs that have been carried out in the name of "eugenics."

With the advances that have occurred both in cellular biology and in tissue culture techniques, human physiological and biochemical genetics have made great strides in the laboratory. Much of molecular biology arises today from studies on human tissue culture cells. Modern cytological techniques have made detailed study of human chromosomes possible. Mouse-human hybrid cells (which tend to lose human chromosomes) offer a splendid method for revealing the chromosomal location of "nutrition-requiring" mutant genes. Such advances, together with the vast amount of familial data that can be stored in and analyzed by modern computers, have once again made the human species an excellent research organism. Within recent years, a multi-billion dollar project has been undertaken with the express goal of obtaining the complete sequence of purine and pyrimidine (nitrogenous) bases in the DNA of the human genome, from the proximal end of chromosome 1 to the distal end of chromosome 23. This project, one of the largest biological research projects ever undertaken, has predictably spun off smaller, test projects such as sequencing the genomes of lower organisms: e.g., *C. elegans* (nematode), *D. melanogaster* (fly), and *A. thaliana* (plant). Indeed, as this paragraph is being written, the complete genome map of a bacterium, *Haemophilus influenza,* has been announced.

This completes the brief (but not all-inclusive) survey of research organisms, the living tools of geneticists. Each was initially chosen because it offered a means for solving certain problems—those pre-existing in the investigator's mind as well as others raised by the organism itself. As the original problems were solved, others took their place. Some of the latter proved to be cul-de-sacs that sit unnoticed even today. Others that might well have been deserted as being unfruitful have been revitalized by the application of new physical tools. Humans, of course, will always be included among research organisms, because many scientists and many legislators (and others with financial power), like Alexander Pope, believe that the proper study of mankind is man.

Physical research tools

Genetics, like other biological sciences, has relied upon mechanical devices from microscopes (the 1000× apochromatic lens was available to cytologists before 1900) to lasers. Geneticists designed some devices, some of which were unsuccessful. Calvin Bridges persuaded Bausch and Lomb to build a compound microscope with the stage as well as the oculars facing the user; this design never caught on. Even the square half-pint milk bottles that Bridges designed (they would not roll off the laboratory bench) were a failure: they blocked the circulation of air in the incubators that were used for raising flies, thus sterilizing or even killing some fly cultures because of the excessively high temperature.

An apocryphal story concerning a preteen Joshua Lederberg, then president of his junior high school science club, states that he and his young colleagues attended a lecture by Robert Chambers at the New York Academy of Sciences. The topic of Chambers's talk was his newly developed micromanipulator. At the conclusion of the lecture, Lederberg's hand was the first one raised in the ensuing question period. "Dr. Chambers," asked the youthful voice, "what contribution do you think your micromanipulator will make to science?" The response, apparently, was not entirely coherent or even persuasive. In many ways, this small vignette presages the plight of national laboratories in which complex and extremely expensive research tools are collected, largely with the hope that someone will eventually find a use for one or more of them.

Magnification

Enlarging objects is always useful in science. The first observers see things no one has ever seen before (Table 3-1). The light microscope, equipped with quartz lenses and a recording film, was adapted for use with ultraviolet light. Because of its short wavelengths, UV microscopy revealed objects with higher resolution than did ordi-

Table 3-1. The resolution capabilities of the eye and of various microscopes. The distances listed are the minimum ones at which two objects are discerned as such rather than as a single object.

	Distance	
	Conventional units	Meters
Eye	0.1 mm	10^{-4}
Light microscope	0.2 μm	2×10^{-7}
UV microscope	0.1 μm	10^{-7}
Transmitting electron microscope	1 Å	10^{-10}
Scanning electron microscope	50 Å	5×10^{-9}

nary light microscopy. The UV absorption characteristics of proteins and DNA added to the uses to which UV microscopy could be put.

Transmitting electron microscopes (TEMs) extended the vision of microscopists immensely. An early version was described (Zworykin, 1941) at the Cold Spring Harbor Symposium on Quantitative Biology. Since then (especially during the 1950s), the electron microscope has revealed to workers the endoplasmic reticulum, the microsomes, the intricate structure (*cristae*) of mitochondria, and the layered structure of chloroplast grana. In the 1980s, transmission electron microscopy provided evidence of the basic unit of chromatin, the association of protein and DNA. These units, called nucleosomes, are composed of eight histone proteins around which is wound DNA.

The purpose of all microscopy is to provide enlarged images of small objects so that what otherwise is invisible becomes visible. The transmitting electron microscope requires that electrons pass through the material being examined; because electrons have poor powers of penetration, the material being observed must be sectioned in extremely thin sheets (40–90 nm; 4–9×10^{-8} m).

The scanning electron microscope (SEM) provides enlarged views of surfaces. It is, for example, extremely useful in revealing the fine details of small morphological structures. Simple organisms such as nematodes reveal details under the SEM that prove to be species diagnostic. Time-consuming, hand-drawn diagrams have been largely replaced by photographs for many taxonomic purposes. Furthermore, when combined with special techniques such as freezing and then fracturing the specimen with a cold microtome knife, the SEM can reveal details of membrane structure or even of such complex organelles as kidney glomeruli.

Separation: Centrifugation

Just as enlargement is useful for a scientific observer, so too is separation—the separation of individual components from a complex mixture. The cream separator serves as an example for those old enough to remember; the cream and milk of "whole" milk were separated by means of a centrifuge. Milk spun to the outside of the spinning rotor and was collected in one outlet; cream "rose"—i.e., was forced to the center of the rotor—and was collected in a second outlet. Useful as early laboratory centrifuges were (and still are in clinical laboratories), the ultracentrifuge represented a major advance in separating large molecules and in estimating their sizes and identifying enzyme or membrane aggregates. Credit for the development of the ultracentrifuge must go largely to a Swedish physical chemist, Theodor Svedberg, who, between 1925 and 1935, improved high-speed centrifuges until they exerted forces 500,000–900,000 times the force of gravity. These forces required rotors that turned at 50,000 revolutions per minute or more—nearly 1000 revolutions per second!

Under these tremendous forces, large molecules (and small organic particles) "spin down." The rate at which the molecules move (see Figure 3-5) allows one to calculate their molecular weight. In the case of particles that consist of characteristic aggregations of smaller molecules (ribosomes, for example), the total mass can be calculated. Thus, ultracentrifugation revealed (in the proper physiological solution) that ribosomes of *E. coli* consist of two parts: a smaller one whose sedimentation rate is 30S (S for Svedberg) and a larger, 50S, ribosome. Combined, their sedimentation rate is 70S. (The S values do not add up because sedimentation is not strictly proportional to mass.) Mammalian ribosomes differ somewhat; their corresponding sedimentation rates are 40S (small), 60S (large), and 80S (combined). Many molecular biologists claim that ultracentrifugation represents the origin of their science. One should remember, however, that *separation* is not *explanation*. Thus, the separation of tobacco mosaic virus into smaller fractions—one protein, the other nucleic acid—was once seen as evidence that nucleic acid is a cement that is merely

Figure 3-5. The sedimentation of the enzyme lysozyme in an ultracentrifuge. Starting at the meniscus (B) separating paraffin oil (A) on the right from water on the left (water is heavier than oil, which "floats"), the lysozyme molecules (C) can be seen "settling out" under a centrifugal force 300,000 times gravity. The bottom of the centrifuge tube is at the extreme left (D). The photographic exposures were taken at half-hour intervals. (Adapted from *The Ultracentrifuge* by Theodor Svedberg and K. O. Pederson [Oxford: Clarendon Press, 1940], by permission of Oxford University Press.)

responsible for protein structure (Svedberg and Pederson, 1940, p. 411).

Contingent upon centrifugal forces and ancillary to the ultracentrifuge are two further refinements, the sucrose gradient and cesium chloride (CsCl) solution. In each case, one achieved by mechanical means (sucrose gradient) and the other by physical characteristics of the salt in solution (CsCl), the centrifuge tube comes to contain a denser suspension fluid at the bottom than at the top. Suspended particles form zones as they spin down through a sucrose gradient of increasing concentration; heavier molecules of the same shape move faster than lighter ones. In a CsCl gradient, under centrifugal force, suspended molecules come to "rest" in layers where their buoyant force matches the force of "gravity" of the surrounding CsCl solution; they sink to that level from above but rise to it from below. Particles of different buoyancy form different (and often clearly discernible) layers under centrifugation in a CsCl gradient.

Separation: Electrophoresis

Substances can be separated not only by centrifugation but also in an electrical field by virtue of their associated electrical charges. A salt solution (e.g., NaCl or, more precisely, Na^+ and Cl^-) transmits an electrical current because Na^+ ions migrate to the negative pole (cathode) while Cl^- ions migrate to the positive pole (anode). Inverted tubes placed over these electrodes collect hydrogen gas at the cathode (atomic sodium displaces hydrogen from water molecules) and chlorine gas at the anode.

Proteins and nucleic acids bear net electrical charges and also migrate in an electrical field, although at a much slower rate than do inorganic ions. The development of *electrophoretic* techniques for the separation of macromolecules is essentially a post–World War II phenomenon, despite pioneering work done in the 1930s (Tiselius, 1937). Two large problems needed solutions before proteins and nucleic acids could be usefully separated. First, the heat generated by the electrical current had to be removed almost instantly; otherwise, for example, proteins would denature and lose their normal biological activity. Crushed ice provided an adequate solution to this problem. Second, the turbulence created by convection currents and the consequent mixing of molecules had to be prevented. The latter problem was solved by the use of stabilizing porous or fibrous matrices such as paper, powders, agar, starch, or polyacrylamide (a synthetic resin). These solutions to the problem of unwanted turbulence cause molecules, especially nucleic acids and proteins, to separate at rates influenced by differences not only in electrical charge but also in mass and shape.

Starch gel electrophoresis (Smithies, 1955) has been especially useful in genetic studies. The apparatus commonly used consists of a shallow Lucite tray, resting on a bed of crushed ice, and filled with a starch gel whose liquid dispersant is the same salt buffer that is contained in small troughs at either end of the tray (and made continu-

DEAD FLY   **WATER-SOLUBLE PROTEINS SPREAD BY MIGRATION IN AN ELECTRICAL FIELD**

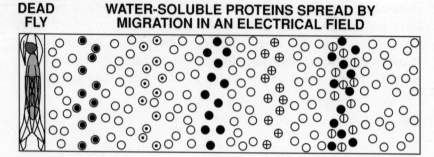

Figure 3-6. A schematic diagram illustrating how the water-soluble proteins in a small organism may be separated by electrophoresis. Having been squashed on a small rectangle of filter paper, the fly (on the paper) is inserted in a narrow vertical slot cut in a thin layer of starch gel. The gel is saturated with a salt buffer whose pH causes most proteins to carry a net negative electrical charge; thus, they move toward the positive pole (right-hand side of the diagram). Electrophoresis provides a mechanism (as does ultracentrifugation) for separating proteins; the problem of revealing the presence of particular proteins is solved by still other techniques. (From *Basic Population Genetics* by Bruce Wallace. Copyright © 1981 by Columbia University Press. Reprinted with permission of the publisher.)

ous with these troughs by buffer-soaked strips of paper toweling). The source of the proteins that are to be separated (the dead and squashed fly of Figure 3-6) is inserted into the gel (usually near the negative pole), and high-voltage electrical current is applied. Ordinarily, the pH of the buffer is adjusted (the electrical charge of a protein molecule is pH dependent) so that most proteins carry a net negative charge and, as a result, migrate toward the positive pole. These proteins, invisible to the naked eye unless they happen to be pigmented, migrate at different rates through the gel and, thus, create a banded proteinaceous smear of some length extending "downstream" from the source—the squashed fly. The separation of proteins carrying different electrical charges is the role of the power source and the electrophoretic apparatus.

As with all technical procedures, modifications have been developed that alter the questions one can ask. The inclusion of a detergent in the gel, for example, disrupts the molecular structure of proteins. The use of a stable pH gradient rather than of a single pH alters the migration patterns of proteins—they come to rest when their net electrical charge is neutral (neither positive nor negative).

X-ray diffraction

At times the structure rather than the composition of a substance is the object of research. Simple enlargement may serve at the cellular level; molecular structure, however, lies far below the resolving power of even the best electron microscopes. The scientist in this case must find a probe that penetrates the molecule much as the custom officer's saber once probed bales of cotton, searching for concealed contraband.

X rays provide the necessary probe; the procedure is known as *X-ray diffraction*. A beam of X rays is aimed at a small crystal of whatever substance is being studied: protein, nucleic acid, or any other molecule. The atoms within the crystal, which are aligned in layers, deflect the X rays onto a photographic film. The dark (i.e., exposed) spots on the film allow the investigator to reconstruct the pattern in which the atoms are arranged. A commonly encountered, comparable phenomenon occurs when one drives on a raised roadway past a regularly spaced, low-lying orchard of small trees. First (if the rows are perpendicular to the highway), one sights down the main, widely spaced rows. Next, upon looking back, one sees secondary and tertiary patterns of more closely spaced rows. If one were provided with the angles at which each of these "layered" patterns became evident, one could easily reconstruct the pattern in which the trees were planted.

In illustrating the use of X-ray diffraction, we cite the early work of W. T. Astbury and F. O. Bell (1938) (see Figure 3-7). Their publication is titled "Some Recent Developments in the X-ray Study of Proteins and Related Structures," and, indeed, the bulk of the paper and the appended discussion concerns proteins. In the course of their investigations, however, the authors examined dried, stretched films of calf nucleic acid (deoxyribonucleic acid, or DNA in today's terminology). Their conclusion, illustrated in Figure 3-7, was that DNA is a fibrous material. In their words (p. 112), "There is the strongly marked period along the fibre axis of 3.34 Å. The natural conclusion is that this spacing is that of a close succession of flat or flattish nucleotides standing out perpendicular to the long axis of the molecule to form a relatively rigid structure." Later, "Roughly, the idea is that of a very tall column of discs with a linking rod down one side."

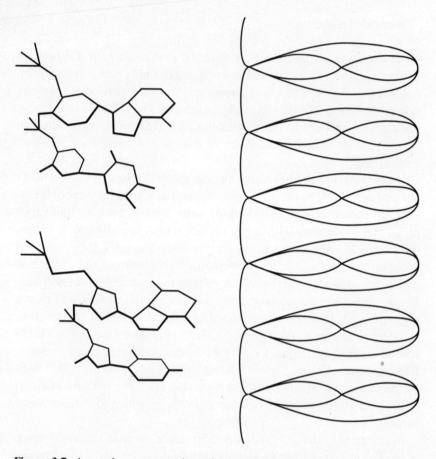

Figure 3-7. An early representation of deoxyribonucleic acid as a column of nucleotides, each of which extends perpendicularly as a flattened disc from the sugar-phosphoric acid fiber that maintains the integrity of the column. The estimated distance between the nucleotide plates was 3.34 Å. At the left are conventional (for the time) chemical representations of the purines, pyrimidines, sugar, and phosphoric acid of "thymonucleic acid"—DNA. (After Astbury and Bell, 1938, by permission of Cold Spring Harbor Laboratory.)

During the discussion that followed Astbury and Bell's report, Stuart Mudd commented (p. 119): "So it seems apparent that these piles of nucleotide units, by slight changes in the order in which nucleotides occur, or possibly by other changes in configuration, give us an adequate basis for specificity." In Mudd's mind at the time, *specificity* meant serological specificity. Astbury, in his reply (p. 119), seems to have had a different (informational) type of specificity in mind:

> There is one point I forgot to mention which bears directly on this [Mudd's] question. You know that the thymonucleic acid is said to be a tetranucleotide [a simple repetitive structure involving the four nucleotides]. We have not worked this out completely, but it is clear from the photographs that the true period along the nucleotide column is at least 17 times the thickness of a nucleotide. So the nucleotides do not follow each other always in the same order, and *this gives you another chance of great variation* [emphasis added].

The account of the eventual modeling of DNA's structure by James D. Watson and Francis H. C. Crick (1953) is too well known to be repeated here; much of Watson and Crick's success lay in X-ray diffraction data obtained by Rosalind Franklin in M. H. F. Wilkin's laboratory. The points to be emphasized here are (1) the considerable information gathered by Astbury and Bell from X-ray diffraction data using extremely crude (noncrystalline) preparations of DNA and (2) the critical importance of *ideas* in the interpretation of observations: the attention of both the authors and the audience at Cold Spring Harbor in 1938 was riveted on proteins, not on this "related structure," DNA. The extent to which conventional wisdom fell short of the truth in 1938 becomes clear upon reading C. H. Waddington's remarks (p. 119):

> Is it not the case that you find the nucleic acid attached to the chromosome at a time when the chromosome is contracted? The only time we know definitely that the nucleic acid is in the chromosome is when the chromosome is still contracting.... Further, one knows that the nucleic acid is not present uniformly throughout the whole chromosome but is concentrated in certain regions, the darkly staining bands or chromomeres.

The necessary framework for a proper interpretation of Astbury and Bell's observations was lacking; Astbury made no substantive reply to Waddington's question, and the next comment (p. 119) reflects the mood of the era: "Going back to the point on globular and fibrous proteins . . ."

In passing, one might note that X rays proved to be valuable in genetic research in ways other than diffraction studies. Once H. J. Muller (1927) had discovered that X rays induce gene mutations, they facilitated the study of the gene, defined as *the unit of mutation;* spontaneous mutations were much too rare to make such a systematic study feasible. Furthermore, many of the physical characteristics of X rays and other ionizing radiations allowed for estimations of the physical sizes of genes.

Other devices

Many more mechanical devices have been developed for use in genetic, especially molecular, research. The laser, for example, is used to induce extremely small lesions in individual chromosomes (Berns et al., 1969); the effects of these molecular injuries can be followed during subsequent embryological development or tissue differentiation. Laser beams can also be focused on individual cells and used to induce color in photoreactive dyes (George Hess, personal communication); cellular communication can then be detected and mapped by the cell-to-cell diffusion of the colored dye. The laser has also led to confocal microscopy, an extremely important tool in developmental genetics. Without the factual information provided by these new, highly sophisticated devices, developmental genetics is largely a science of the *possible* rather than one of the *actual;* an excellent example of a plausible but erroneous assumption appears later (p. 171) with respect to an eight-cell structure in the *Drosophila* eye.

Confocal microscopy, because it involves several technical advances, deserves special comment. The product of confocal microscopy is an extremely high-resolution image of a section of tissue (or, for example, of a whole-mounted insect's egg) in which many individual proteins are identifiable by color, in situ. This precision has been made possible because each protein has been bound to a protein-specific antibody, which, in turn, has been labeled with a fluorescing molecule of a given color. The following technical procedures were needed in order to achieve these spectacular results:

- *Monoclonal antibody,* an antibody produced by a clone of cells that are descended from a *single* immune cell of a rabbit (or other mammal). This antibody binds with a specific protein, and no other. The original immune cell has been "immortalized" by fusion with a cancer cell; the specific (hence, *mono-*) antibody is produced in large quantities by cultures of these fused cells (also known as *hybridomas*).

- *Fluorescent labels* are molecules that emit light of various wavelengths when illuminated by a laser beam. These labels were once tagged onto antibodies for antibodies but are now often tagged onto the monoclonal antibodies themselves.
- A *laser beam* is focused on an extremely thin plane within the specimen (a plane that is much thinner than the thinnest possible microtome section; hence, the high resolution). This plane can be raised or lowered at will, thereby allowing observations at different levels within the specimen. The fluorescently tagged antibodies reveal where specific proteins are located within the section. (A splendid example of confocal microscopy can be found in Ripoche et al., 1994; theirs was a study of the location of Golgi membrane proteins in *Drosophila* embryos.)

Many of the devices one encounters are, of course, merely labor-saving devices. Science consists largely of routine tasks—blending, separating, warming, cooling, and others—that can be mechanized. Aside from these mechanical aids, however, are devices of the sorts described here that perform tasks human beings cannot perform by themselves—gadgets that allow us to see, to separate, and to probe. These gadgets, under the guidance of ideas or hypotheses, lead to scientific discoveries. Indeed, there is a branching point that characterizes each technical innovation (Figure 1-1): the application of the technique to already existing problems and the generation of new ideas that could not have arisen prior to the technical advance.

Searching, locating, and revealing

Much of science consists of making and verifying predictions. At times, as in the case of nondisjunction in *Drosophila melanogaster* (Bridges, 1916), the process is reversed: rare exceptions are observed, and testable explanations are formulated as a consequence. The relatively high frequency of wild-type revertants that occur within stocks of *Bar*-eye mutants (as well as the corresponding occurrence of *Double-bar* eyes) also demanded an explanation, one that was provided (and confirmed by cytological examination) by Calvin Bridges in 1936. The *Bar*-eye phenotype is caused by a small, tandemly oriented chromosomal duplication: . . . *abcdabcd* . . . During oogenesis when homologous chromosomes pair, two pairing patterns are possible in the case of homozygous *Bar* females:

$$\ldots abcdabcd \ldots \qquad\qquad \ldots abcdabcd \ldots$$
$$(1) \ldots abcdabcd \ldots \qquad (2) \ldots abcdabcd \ldots$$

In the case of "normal" pairing (1), a recombination occurring anywhere within the illustrated region leaves the resulting chromosome strands unchanged. On the contrary, a recombination that occurs within the paired segment shown in (2) results in two dissimilar daughter chromosomes. One of these can be represented ... *abcd* ...; that is the normal (wild-type) pattern. The second pattern, ... *abcdabcdabcd* ..., exaggerates the *Bar* phenotype that results from the duplication of this small chromosomal segment; the extremely small eye caused by the triplication of the abcd-segment was named *Double-Bar*.

Making rare events common

Encountering rare exceptions is one thing (these are occasions that, according to folklore, favor the prepared mind); predicting their occurrence and then finding them is quite another. Muller's use of chromosomal inversions that lock many gene loci together so that they are inherited as a single unit provides an excellent example of the skillful manipulation of one's research material. Suppose that in *Drosophila* a given sex-linked gene, *A*, mutates to a recessive lethal, *a*, at a rate of 10^{-5} per generation. Suppose further that one can set up matings such that the absence of *a*Y males can be noted. Finally, suppose that an experimental treatment (X-radiation, for example) increases the mutation rate 10-fold. The last point is the one that is to be demonstrated. A convincing demonstration requires several spontaneous mutations (say, six) in order to give a reasonable measure of the spontaneous mutation rate and 60 or so mutations in the treated series. The total number of tests needed to obtain such convincing data equals $600,000 \times 2$, or 1,200,000 tests in all. If one person could perform 1000 tests per day, the entire experiment would require 1200 days, or more than three years.

By using chromosomal inversions that conserve blocks of genes (about 300 gene loci for the X chromosome of *D. melanogaster*) as single units, Muller dealt with a spontaneous mutation rate equal to 300×10^{-5}, or 0.003. A radiation treatment that increased mutation rates 10-fold would result in an induced mutation rate of 0.030. Thus, in order to obtain six spontaneous and 60 induced mutations, one need run a total of only 2000 tests for each (control and experimental) series—4000 tests in all. Anyone who could perform 1000 tests per day would require only four days—not three years—to complete the study. Although this exercise has been presented as an abstraction, it has a basis in reality: T. H. Morgan (1914) tested radium rays for their possible mutagenic effect, but with negative results. He lacked Muller's carefully designed experimental materials and procedures.

In one sense, the use of chromosomal inversions for tying together large numbers of gene loci is a statistical ploy that increases one's sample size. Without dwelling on the matter at length, we note that mathematical, often statistical, reasoning underlies many advances in genetics. This was certainly true in the case of Mendel's analysis of the results he obtained by crossing different varieties of the garden pea. It is equally true in the case of Luria and Delbrück's (1943) demonstration that mutant individuals exist in bacterial cultures prior to the test that reveals their presence. If the mutant individuals arise (á la Lamarck) in response to a given challenge, they must do so with some low probability. Consequently, the numbers of mutant individuals among independent cultures should follow the *Poisson distribution:* the variance of the number of mutant bacterial colonies appearing on different culture (petri) dishes should equal the average number of colonies per dish. Luria and Delbrück showed, however, that the variance among cultures was much greater than the mean, thus revealing a second source of variation, namely, the preexistence of mutant individuals in the initial (small) inocula plus the random times at which such mutants subsequently arose during the growth of each culture. Lederberg and Lederberg (1952) demonstrated the existence of these mutant bacteria in a most elegant physical manner, by *replica plating.* A sterile velvet disc was pressed down on a petri dish on which many small bacterial colonies were forming. This disc, with the many bacteria from each colony clinging to its nap, was then pressed against the agar surface of one or more additional petri dishes; in each case a pattern of colonies was formed corresponding exactly to that of the original plate. The daughter colonies could be tested for antibiotic or phage resistance (or for any other mutant trait). In each instance in which a mutant colony was discovered, the original plate had the corresponding mutant colony in the same position. Mutations, consequently, do not arise in response to a challenge but, rather, preexist and are merely revealed by an appropriate challenge. (This view has recently been challenged itself, largely by the alternative view that stressful conditions increase an organism's mutation rate—i.e., the rate at which it makes genetic errors.)

Revealing individual proteins after electrophoresis

In the previous section, electrophoresis was described as a means by which substances (especially proteins and nucleic acids) can be separated from one another. If these substances are pigmented (e.g., hemoglobin), their separation can be followed visually. If they are not pigmented but are plentiful, a nonspecific protein stain (e.g., Coomassie Blue or picric acid) may reveal what otherwise cannot be seen. On the other hand, if, as in the case of enzymes, the substance exists in extremely low concentrations, then it must be forced to reveal itself. Starch gel electrophoresis (Smithies, 1955) profited greatly from techniques that had been developed earlier by histochemists. If the substrate (A) that a given enzyme can degrade into "daughter" substances B and C is known, it is often possible to reveal the presence of one of these products, say by its interaction with a dye: B + (leuco) dye $\rightarrow$ D + dye (both colored and insoluble). The position of the enzyme is thus revealed on the starch gel by the formation of a colored band. This technique can be expanded by the addition of an appropriate, second enzyme to the "developer" solution. Thus, the enzyme whose presence is to be revealed degrades A into B and $C;$ the enzyme that has been added to the developer solution degrades C into D and $E;$ E then reacts with the leuco-dye to form an insoluble colored band. The analyses of starch gels comprise many variations on the same general theme, including the detection of amylases by the digestion of the starch gel itself. In each instance, the distance separating different molecules possessing similar enzymatic actions must be large relative to the distances these molecules (and those of the insoluble dye) can diffuse in the starch gel. The histochemical techniques were largely unsuited for their originally intended purposes because histological (cellular) distances are so small.

The matrix—paper, starch gel, agar, synthetic polymer, etc.—within which electrophoresis is carried out may be suitable for one purpose (e.g., revealing the presence of a specific enzyme) but not for another (e.g., serological testing or nucleic acid hybridization). The material on one gel needs to be transferred intact to a second one. For proteins, the necessary transfer can be accomplished by electrophoresis! Empty gels of various sorts are stacked atop the original one with its dispersed material, like playing cards, and the electrical current is passed vertically through the deck. If all cards, to continue the analogy, were blank except for an eight of hearts at the bottom, the electrical current would cause the red pigment to migrate upward through the blank cards, thus transforming each of them into a (somewhat pale) eight of hearts. Each so-called blot could then be subjected to an appropriate further analysis (see p. 44).

Radioactive probes

DNA and RNA offer still other opportunities for detection because of the tendency for complementary single strands to form double-stranded molecules (DNA-DNA, DNA-RNA, or RNA-RNA) when offered an opportunity. Thus, to locate a particular fragment of DNA among many that have been enzymatically generated and that have migrated to various positions on an electrophoretic gel, one only need to flood the gel (or its corresponding "blot") with radioactively labeled complementary DNA (or RNA), incubate under appropriate physical conditions (temperature, salt concentration, and the like), rinse, and locate the radioactively labeled band by exposing the gel to an X-ray film. This and related procedures are dealt with in the subsequent section on genetic engineering.

The use of radioactive isotopes in physiological studies, although originating in the 1920s, really stems from work on photosynthesis in the late 1930s. An abundance of radioisotopes of many sorts became available only after World War II. Because the *chemical* properties of elemental isotypes are identical, beta- or gamma-ray emissions of a radioactive isotope reveal the fate of that element in chemical reactions. An example of the use of these atomic tracers is provided by the study of photosynthesis. The elementary textbook basic formula is given as

$$6CO_2 + 6H_2O \rightarrow C_6H_{12}O_6 + 6O_2.$$

Thus, one might believe that ^{14}C-labeled $^{14}CO_2$ would be found in the glucose molecule: $^{14}C_6H_{12}O_6$. Not so! A five-second exposure of green algae to $^{14}CO_2$ and light produces radioactively labeled 3-phosphoglyceric acid (PGA). The acceptor of CO_2 is ribulose diphosphate (RDP); thus, the first (chemical) step in photosynthesis can be written

$$RDP + CO_2 \rightarrow 2 \ PGA.$$

Neither glucose nor water is involved in these initial reactions.

With the development of bacterial and phage genetics, it became necessary for experimenters to locate small needles in very large haystacks—one, two, or three exceptional bacteria among tens of billions of ordinary, unwanted ones. To a large extent, these searches were done by genetic means. Lederberg and Tatum (1946), for example, demonstrated that *E. coli* undergo genetic recombination by mixing two strains on a minimal medium that was incapable of supporting the growth of either one. The two strains can be represented as *abCD* and *ABcd;* for growth, each required that the minimal medium be supplemented by two nutritional additives. Furthermore, the two requirements of the two strains differed. Most of the plated bacteria did, in fact, fail to produce colonies (i.e., failed to grow and reproduce); however, a few colonies did appear on these plates. The bacteria constituting these few colonies proved to carry a normal allele at each of the four gene loci: *ABCD.* This was possible only if individual bacteria had access to genes carried by individuals of the other strain and then substituted normal alleles for their own mutant ones: *abCD(+AB)* → *ABCD,* or *ABcd(+CD)* → *ABCD.* The ability to make novel combinations from genes that are contributed by two separate individuals (such as husband and wife) is the essence of sexual reproduction. Hence, Lederberg and Tatum demonstrated sexual reproduction in bacteria, a demonstration that allowed these organisms to be subject thereafter to standard genetic analyses.

Progress in bacterial genetics has frequently relied upon selective procedures that allow only wanted or expected (but rare) individuals to grow or that allow these rare individuals to identify themselves on "indicator" (dye-forming) plates. Isolating bacterial strains that are resistant to antibiotics is easy; one plates many bacteria on the antibiotic-containing medium and saves the surviving colonies. And what about selecting for biochemically deficient bacteria? With the aid of penicillin, it is not difficult. Bacteria are placed in a culture tube containing minimal medium (a medium on which only wild-type strains can grow) to which penicillin has been added. Because penicillin kills only growing and dividing cells, those bacteria that cannot grow on minimal medium (i.e., those that are nutritionally deficient) are immune to its lethal action. One then rescues the few surviving individuals; these are the ones that were unable to grow because of one or another nutritional requirement (Davis, 1948). Note what experimenters do in performing these experiments: They specify the event they wish to recognize. They design the experiment so that all unwanted events are ignored. And, because the event (mutation or recombination, for example) is manifest in a living organism, it and its descendants can be isolated, perpetuated, and grown in huge quantities in the laboratory.

Genetic engineering by recombinant DNA technology

The term *genetic engineering* as it is used here includes the prerequisite ability to manipulate macromolecules in vitro without destroying their biological function. (Of course, plasmids, bacteriophage, and other viruses that move host cell DNA from one cell to another carried out "engineering" feats long before human beings evolved.) Under our definition, one can argue that genetic engineering began with the crystallization of tobacco mosaic virus (Stanley, 1935) and the subsequent demonstration that, when dissolved, the virus remained infective. Alternatively, one might cite the work of O. T. Avery, C. M. MacLeod, and Maclyn McCarty (1944) in which pneumococci (*Streptococcus pneumoniae*) were fractionated into protein, polysaccharides, and nucleic acids followed by the demonstration that the DNA could alter the genetic characteristics of recipient cells. Harriet Ephrussi-Taylor (1951) summarized the available information on bacterial transformation, stressing that the free DNA becomes incorporated into the bacterium's genetic material by recombination. Another approach to genetic engineering—plasmid and chromosome transfer—represents the harnessing and conscious directing of the ability already possessed by extrachromosomal DNA elements (called plasmids).

Macromolecules, especially proteins and nucleic acids, are rather delicate with respect to manipulations that allow both for their isolation and for their retention of normal (i.e., biological) activity. Enzymes are inactivated by heat and by acids, heavy metals, proteolytic enzymes, other chemicals, and by the loss of essential cofactors such as vitamins. The three-dimensional conformation of a protein, not merely its amino acid sequence, is necessary for it to function properly. Destroy its conformation and the biological activity of the protein is also destroyed. Heat, enzymes, and other substances can also destroy the biological activity of DNA. During the 1960s and 1970s, procedures were developed that allowed one to isolate and produce electrophotomicrographs of entire complements of DNA from phage (1962) and bacteria (1963). Subsequently, about two-thirds of the DNA in one of *D. melanogaster*'s large, V-shaped chromosomes (molecular weight, 2.8×10^{10}) was isolated as a single, truly gigantic molecule.

The discovery of *restriction enzymes* was a landmark in the history of molecular genetics and genetic engineering. Complementary strands of DNA pair with one another spontaneously. Within DNA molecules, there are regions that correspond to reverse repeats ("palindromes"; recall "Able was I ere I saw Elba"):

$$-\text{ATTCGGCCGAAT} \rightarrow$$
$$\leftarrow \text{TAAGCCGGCTTA}-$$

The arrows indicate that each strand of DNA has a polarity that reflects its molecular structure; the polarities of the complementary strands of double-stranded DNA point in opposite directions as the arrows indicate. Restriction enzymes commonly cause staggered cuts in the two strands of DNA, two or three bases removed from the point of symmetry in a palindromic sequence:

$$-\text{ATTC} \rightarrow \qquad\qquad -\text{GGCCGAAT} \rightarrow$$
$$\leftarrow \text{TAAGCCGG}- \qquad\qquad \leftarrow \text{CTTA}-$$

The resulting fragments can be seen to possess "sticky" ends that, if given an opportunity, will rejoin to form an intact DNA molecule once more. However, the sticky ends produced anywhere within a DNA molecule by a given restriction enzyme are identical because the enzyme recognizes and cuts the same palindromic sequence wherever it occurs in any DNA fragment, whatever its source. Hence, a mixture of DNAs from diverse sources, when cut with a given restriction enzyme, leads to the joining of the DNA from those different sources.

	...ATTC	GGCCGAAT...	
fly DNA			bacterial DNA
	...TAAGCCGG	CTTA...	

Herein lies the genetic engineer's ability to create genes and gene products that are impossible to find in the natural world.

During the 1970s, a number of chemical advances occurred that greatly advanced the molecular geneticist's ability to manipulate DNA. One of these was the independent discovery by Howard Temin and David Baltimore in 1970 of *reverse transcriptase,* an enzyme useful in experimental protocols. Reverse transcriptase synthesizes DNA from a single-stranded substrate RNA; it is coded for by retroviruses—RNA viruses—and permits these viruses to make DNA molecules that correspond to their own RNA genomes. The DNA "copy" of a retrovirus can be incorporated into the host's genome. At any time after incorporation, the DNA in the host's genome can be transcribed to generate the RNA virus anew: the virus need no longer penetrate the cell from the outside because the directions for its synthesis are now carried by the host! In higher organisms, the DNA that specifies a particular gene product often carries intervening segments of DNA ("introns") that must be removed during or immediately after the transcription of mRNA; otherwise the amino acid content of the bacterial synthesized protein product would be wrong and the protein would be nonfunctional. Hence, reverse transcriptase is used by genetic engineers to make DNA that corresponds to an actual mRNA (messenger RNA) molecule; this copy DNA (cDNA) can be inserted into bacteria in a way that allows them to generate mRNA and, from that mRNA, the protein that it specifies.

Advances in the technology of genetic engineering are occurring so rapidly that it is fruitless to attempt here a complete catalogue. Such a catalogue, like many engineering handbooks, is best kept by serious practitioners in a loose-leaf notebook, with new leaves inserted daily and old ones removed. The up-to-date information is best handled by computer-based exchanges with electronic databases.

This account of genetic engineering techniques can conclude with an expanded account of three specific blotting procedures: the Southern blot, the Northern blot, and the Western blot. (Only *Southern* is a person's name; the other two "directions" derive from the humor of molecular biologists.) Each of these blots is essentially a refined derivative of the Lederbergs' replica plating technique. After DNA fragments (Southern), RNAs (Northern), or protein molecules (Western) have been separated in an electrophoretic gel, a moist special paper membrane is placed on the gel. The molecules within the gel beneath the paper are drawn into the paper either by capillary action or by an electrical current. The molecules transferred to the paper can then be probed (to "probe" is to apply a technique that reveals the sought-for molecule or molecular fragment) by DNA or RNA (Southern or Northern) or by chemically tagged antibodies (Western). The nucleic acid probes may be labeled with radioactive phosphorus (^{32}P) or a nonradioactive (chemical or enzymatic) tag. The upshot is that the locations of specific nucleic acids or proteins on an otherwise undecipherable gel are clearly revealed. The general problem, once more, is one of revealing where molecules of interest are to be found: to locate the very small needles in a very large haystack.

DNA that has been synthesized by the action of reverse transcriptase on a particular mRNA can be used as an in situ probe of the existence and spatial distribution of that mRNA in tissues and, especially, in developing embryos. As one might expect, DNA probes reveal the presence of the DNA's complementary mRNA before the presence of the corresponding protein can be detected (e.g., by confocal microscopy).

Advanced chemistry

An important coincidence occurred during the early 1950s: the Watson-Crick model of DNA structure was developed simultaneously with the realization that all molecules of a given protein (hemoglobin, ribonuclease, or insulin, for example) have identical amino acid sequences—not merely the same amino acid composition. The latter was a fact long known to physiologists. Each protein molecule is constructed with amazing accuracy. The Watson-Crick model for DNA suggested at once (1) how DNA is replicated with its own incredible accuracy and (2) how the sequence of nitrogenous bases in the DNA molecule might control, or specify, the sequence of amino acids in protein molecules (recall the account that accompanies Figure 3-7). By the mid-1960s, the genetic code had been deciphered; the sequences of three bases (codons) that either specify each of the 20 amino acids or ("stop" codons) halt the growth of polypeptide chains were known.

Lacking for many subsequent years was a means by which the sequence of bases in an isolated fragment of DNA could be ascertained. Rather laborious procedures for identifying the sequence of amino acids in a given polypeptide were available, but, because of the redundancy in the genetic code (61 codons specify 20 amino acids), information about an amino acid sequence cannot provide firm evidence concerning the DNA that is responsible for its synthesis. For example, take the amino acid sequence

Tyr-Trp-Pro-Leu-Ser

The number of codons specifying each of these five amino acids, respectively, is 2, 1, 4, 6, and 4. Thus, 192 different nucleotide sequences in DNA could specify this particular sequence of five amino acids.

The magnification of DNA molecules, even with the use of the electron microscope, has failed to yield any information regarding the nucleotide sequences in isolated DNA fragments. Chemical methods (either cleaving DNA at specific bases or terminating the synthesis of DNA at specific bases) have succeeded, however.

Here, one such method is described sketchily (based on Sanger et al., 1977):

With a single strand of DNA of *one* (pure) sort available, one can, by the use of a DNA polymerase and a primer, and in the presence of ample amounts of all four nitrogenous bases, synthesize numerous complementary copies: First a thymidine (for example) starts the new strand, then an adenine is added, then a thymidine, then a guanine; all the while, the new strand gets longer and longer. When the complementary strand matches the original template strand in length, growth ceases.

If to the mixture of nitrogenous bases, a small quantity of *di*deoxyribonucleotides is added, the growing strand will terminate at whatever point this dideoxy compound is inserted because a *di*deoxyribonucleotide acid lacks the hydroxyl group that reacts with the phosphorus of the following (next-to-be-added) nucleotide. Four terminator dideoxys are available: ddATP, ddTTP, ddGTP, and ddCTP. When appropriate amounts are added to each of four reaction tubes, one of the four terminators to each tube, there is a small probability that a dideoxy-terminator rather than a deoxyribonucleotide will be inserted in the growing DNA molecule.

Consider a small portion of a single-stranded DNA template:

...TGACT...

In the presence of a terminating ddATP (the terminating adenine is represented as A*), some complementary chains will terminate with an early insertion (A*); others will terminate at the fifth insertion: ACTCA*. In the presence of ddCTP (a second reaction tube), the terminated complementary strand will be AC*. With ddGTP, it will be ACTG*. And, with ddTTP, ACT*.

High-resolution electrophoresis clearly separates fragments of DNA that differ in length by a single nucleotide; thus, by electrophoresing the contents of the four tubes in adjacent lanes on a polyacrylamide gel, one obtains the following pattern:

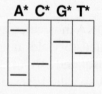

Starting at the bottom with the small, fast-moving fragment, and reading upward over all four channels, one can reconstruct the sequence of bases in the complementary strand as

...ACTGA...

Notice that this strand does, indeed, complement the one originally postulated. The laboratory (enzymatic DNA synthesis) procedure described here is so convenient that enormous lengths of DNA cloned from many sources have now been sequenced. The resulting sequences can then be scanned (by eye or by computer) to see what proteins, if any, their RNA transcripts might specify. Three codons of 64 are "stop" codons; they terminate the growing polypeptide chain. Thus, if a sequence of nucleotides in a DNA fragment is read, starting with the wrong base (i.e., corresponding to a frameshift mutation), a stop codon is likely to be encountered within the first 15–30 (average = 64/3) codons. If, however, one encounters a stretch of 100–300 consecutive codons that starts with a methionine (AUG, or "start") codon and continues with each succeeding codon specifying an amino acid (such a region is known as an "open reading frame"), one can be reasonably sure that this region of the DNA does, in fact, specify the sequence of amino acids for a particular protein—a protein whose function may or may not be known.

With technical skills now available to molecular biologists, mRNA can be isolated from individual cells, the mRNA can be transcribed as DNA by reverse transcriptase, the DNA can be (1) cloned (requires restriction enzymes), (2) sequenced, (3) amplified in quantity by the *polymerase chain reaction* (PCR; described in Chapter 10), and (4) expressed in the synthesis of the protein itself.

Solid phase synthesis

To the skills tabulated above, one must now add the chemist's ability to synthesize both proteins and DNAs *to any specification.* In both procedures, the starting element (amino acid in the case of a polypeptide; nitrogeneous base in the case of DNA) is attached to a small, inert particle; that step prevents the simultaneous growth of the linear molecule in both directions. The solution containing the growing molecule is exposed stepwise to the sequence of elements that are to be added. In fact, the synthesis of tailor-made macromolecules is now automated.

Summary

After this lengthy account of experimental materials and techniques, it seems appropriate to state clearly what has not been intended: no one has been expected to memorize the cited examples as small factoids orbiting in an otherwise empty space called "science." Indeed, laboratory jargon—so necessary for those involved in molecular research—has been shunned in this account. What *is* intended is that research organisms, research tools, and investigative techniques be recognized as providing the means by which puzzling questions are solved. The tools and techniques have been developed over considerable spans of time (despite the modernity of Garrod's biochemical investigations of 1909). Viewed in that light, this chapter has described some of the intellectual and technical advances arrayed on the sound board of the harp shown in Figure 1-1; each was applied to preexisting problems, but each, in turn, gave rise to new problems that awaited still newer advances.

4 Morphology

Embryologists and geneticists have long sustained a most durable of all biological controversies. Embryology preceded genetics as a science. Thomas Hunt Morgan was an experimental embryologist before he turned to the genetics of *Drosophila melanogaster*. Ernst Boveri was a superb experimental embryologist. The factual matter with which all embryologists were (and still are) confronted is the following: a fertilized egg, by means of repeated cell divisions, is transformed into a small mass of dissimilar cells, then into a collection of tissues, and finally into the aggregation of organs and organ systems that constitute the individual. Actually, there are two matters. First, the undifferentiated egg, in a seemingly miraculous manner, gives rise to a functioning individual that is appropriate for the species; frogs develop from frogs' eggs, dogs from those of parental dogs, and human beings from human beings. Second, however, is the variation that occurs within the species. Except at night, all cats are not gray; they are white, black, red, calico, blotched, solid colored, tabby striped, short haired, long haired, and much more. Thus, intraspecific (and other) variation poses a problem that lies atop the larger one concerning the individual's normal development.

Early geneticists, following the dicta set down by their cytologist colleagues, emphasized the constancy—or, at least, the accuracy—with which chromosomal material is apportioned in strictly equivalent amounts between daughter cells. This point-by-point apportionment had been noted and correctly interpreted by Wilhelm Roux in 1883.

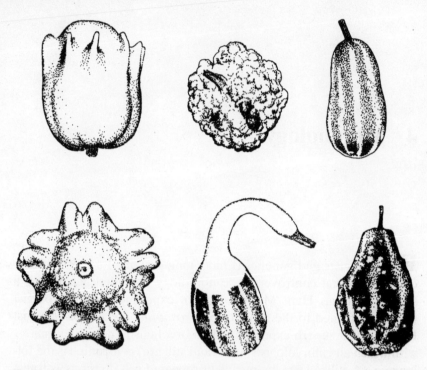

Figure 4-1. A collection of variously shaped and colored gourds. The different patterns of color and the often bizarre shapes are constant enough to form the bases for identifying species and subspecies. The abrupt changes in either color or shape pose excellent questions for developmental genticists.

The fundamental conflict between embryologists and geneticists, then, has involved the origin of cellular and tissue variation from a notoriously constant genetic program. The predictability of genetic variation, even that which alters an individual's morphology, was admitted by even the most committed developmental biologist; this variation was largely regarded, however, as superficial and trivial (see Figure 4-1). The normal differentiation of tissues and organs was (and to some extent still is) regarded as epigenetic—beyond genetic control. Accordingly, John A. Moore (1963) concluded his book *Heredity and Development* with a chapter titled "Developmental Control of Genetic Systems." Perhaps, by the end of the present chapter, the interplay of genetic programs and cellular differentiation will become clear.

Mendel and other early geneticists, intent as they were on demonstrating the *patterns* of heritance, followed characteristics that were easily seen. In part, these were differences in color, a matter that involves the chemistry of pigments; this matter will be covered in Chapter 5. The remaining traits were morphological—traits that involve shapes, sizes, and segmentation patterns.

Mendel's classical analyses of inheritance in peas, *Pisum sativum,* involved contrasting states of these characters:

- form of ripe seeds
- color of seeds
- color of seed coat
- form of ripe pods
- color of unripe pods
- position of flowers
- length of stem

The color and shape of seeds and other plant tissues were markers that allowed Mendel to follow the inheritance pattern of a to-him-invisible material substance. How this substance functioned lay beyond his investigative powers. He was content to make these comments (Mendel, 1865): "In the opinion of renowned physiologists, for the purpose of propagation one pollen cell and one egg unite . . . into a single cell, which is capable by assimilation and formation of new cells to become an independent organism. This development follows a constant law, which is founded on the *material composition* and *arrangement of the elements* which meet in the cell in a vivifying union" [emphasis added].

Although the differing morphologies that Mendel used to such great advantage are frequently cited in the teaching of genetic ratios to young students, relatively little has been done in attempting to understand their physical basis. Relatively little, but *not* absolutely nothing. Consider, for example, the length of the stem. Mendel (1865) noted that the dwarf × tall hybrids were taller than the tall parent: "It must be stated that the longer of the two parental stems is usually exceeded by the hybrid. . . . Thus, for instance, in repeated experiments, stems of 1 ft. and 6 ft. in length yielded without exception hybrids which varied in length between 6 ft. and $7\frac{1}{2}$ ft." That statement of fact has scarcely been extended during the intervening 130

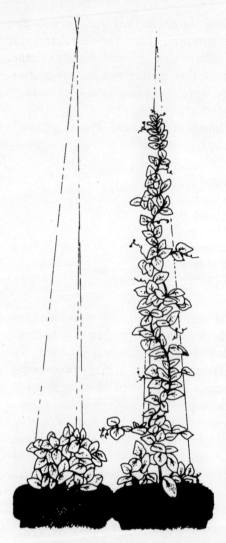

Figure 4-2. The "conversion" of a dwarf pea plant by exposure to gibberellin into a plant whose height corresponds to that exhibited by individuals of a tall variety. In his studies, Mendel determined the pattern of inheritance for the two contrasting morphologies; today's geneticist knows that the morphologies reflect differences in the biochemical genetics of the pea varieties.

years. The gene action responsible for this phenomenon, called hybrid vigor, remains largely unknown.

More progress has been made regarding the difference in height between dwarf and tall pea plants. Figure 4-2 illustrates the effect of gibberellin (a plant growth hormone) on the growth of dwarf peas. The typical dwarf plant is shown on the left; a dwarf plant that has been treated with gibberellin is shown on the right. The treated dwarf attains the height of a normally tall pea plant.

One might assume, then, that a dwarf pea is deficient in gibberellin (or gibberellic acid). Not so! Extracts of dwarf peas reveal that they contain gibberellin(s) in quantities comparable to those of the tall varieties. Thus, one may conclude that the short stature of at least some dwarf varieties of peas may involve the presence of a gibberellin antagonist rather than a lack of this growth hormone (Corcoran, 1976).

Much of the research on dwarf peas has been concentrated on unraveling the metabolic pathways by which a dozen or more gibberellins are synthesized. Thus, the dwarf peas studied by Mendel have normal levels of some gibberellins but lack others; it is only in the presence of those others that normal (i.e., tall) growth of dwarfs occurs (Figure 4-2). The role gibberelins play (as classic growth hormones) in regulating gene action is still unclear. First, relevant studies involve many different plant species in which events follow diverse paths. Second, the techniques required for answering precise questions regarding gene action have been developed only within the last decade or so. Consequently, one can say scarcely more than that the treatment of dwarf pea seedlings with gibberellin results in an increased synthesis of DNA and RNA. One might ask: messenger RNA as well as other RNAs? Or, messenger RNA for specific gene loci? Transcription of new RNA or merely lowered turnover rates? These seem to be matters of considerable uncertainty with respect to gibberellins; greater progress has been made in studies of other plant hormones.

The *bithorax* mutation in *Drosophila melanogaster:* A case study

Many mutant genes in *Drosophila melanogaster* are "visible" mutations; that is, they alter the fly's phenotype. Some better-known ones are listed in Table 4-1. Beginning with the classic sex-linked white-eye (*white, w*) mutant that was discovered in May 1910, the list extends through the first decade of *Drosophila* genetics. Among these mutants are some that affect pigmentation (white eyes, black body, and sepia eyes, for example). A great deal is known of the metabolic

Table 4-1. Dates when some well-known mutants of *Drosophila melanogaster* were discovered.

Mutant	Obvious effect	Date discovered	Discoverer
white (*w*)	Eye color	May 1910	T. H. Morgan
black (*b*)	Body color	Oct. 1910	T. H. Morgan
vermillion (*v*)	Eye color	Nov. 1910	T. H. Morgan
vestigial (*vg*)	Wing shape	Dec. 1910	T. H. Morgan
yellow (*y*)	Body color	Jan. 1911	E. M. Wallace
ebony (*e*)	Body color	Feb. 1912	E. M. Wallace
forked (*f*)	Bristle shape	Nov. 1912	C. B. Bridges
Bar (*B*)	Eye shape	Feb. 1913	S. C. Tice
sepia (*s*)	Eye color	May 1913	E. M. Wallace
bithorax (*bx*)	Morphology	Sept. 1915	C. B. Bridges
cut (*ct*)	Wing shape	Oct. 1915	C. B. Bridges
scute (*sc*)	Bristle (missing)	Jan. 1916	C. B. Bridges
scarlet (*st*)	Eye color	Nov. 1916	M. H. Richards
singed (*sn*)	Bristle shape	Oct. 1918	O. L. Mohr
brown (*bw*)	Eye color	Oct. 1919	G. H. M. Waaler
crossveinless (*cv*)	Wing vein	Dec. 1919	C. B. Bridges
cinnabar (*cn*)	Eye color	Sept. 1920	R. E. Clausen

pathways that lead to these altered pigments. Those involving melanin are especially well known, as one might surmise from the early work of Garrod (Chapter 2) and subsequent studies by Beadle and Tatum (1941a, b).

Several of the mutants listed in Table 4-1 affect the morphology of the fly. *Vestigial* (*vg*) wings, for example, are smaller than normal, wild-type wings, even to the point of being virtually absent. *Forked* (*f*) and *singed* (*sn*) bristles are grotesquely misshapen relative to the long, smoothly tapered bristles of a normal fly. *Cut* (*ct*) and *crossveinless* (*cv*) wings exhibit seemingly minor abnormalities; the border of the wing is scalloped in one case, and the crossveins are interrupted or missing in the other.

Bithorax (*bx*), discovered in 1915 by Calvin B. Bridges, was the first *homeotic* mutant to be noticed in *Drosophila melanogaster*. A homeotic mutation is one in which one appendage or organ of an insect's body is altered so that it resembles another. The names of some such mutants in *D. melanogaster* reveal the changes that have been observed: *aristapedia* (the arista—i.e., the terminal portion of the fly's antenna—takes on the appearance of a leg), *proboscipedia* (mouth parts appear leg- or arista-like), *Polycomb* (sex combs occur on the second and third legs of the male fly in addition to those normally present on the anterior legs), and *Hexapter* (a winglike appendage on either side of the prothorax). (In *Drosophila,* as in all insects, the adult thorax arises from a fusion of three larval segments that constitute the pro-, meso-, and metathorax of the adult fly.) In the *bithorax* mutant, the anterior half of the metathorax becomes mesothoracic. The posterior half remains unchanged. The haltere, an appendage of the metathorax that is homologous to the hind wing of a four-winged insect such as a bee, butterfly, beetle, or dragonfly, becomes winglike, thus, resembling the wing normally found on the mesothorax. Referring to *Hexapter* once more: Some extinct dragonflies of the Carboniferous period possessed three pairs of wings; the first pair was borne by the prothoracic segment just as in the case of hexapterous *D. melanogaster.*

The earliest literature dealing with *bithorax* dealt, of course, with its discovery and with the by-then standard genetic analyses that led to its location (map position 58.8 on the third chromosome, one of *D. melanogaster's* two, large, V-shaped chromosomes). These formal analyses continued as more *bithorax*-like mutants were discovered: bx^3, 1925; bx^{34e}, 1934; bx^w 1934; *bxd* (*bithoraxoid*), 1919; *Cbx* (*Contrabithorax*), 1949; *pbx* (*postbithorax*), ca. 1954; and *Ubx* (*Ultrabithorax*), 1934. These mutant alleles, when located within the fly's genome, all mapped to the same region of chromosome 3.

Edward B. Lewis,[*] one of Alfred H. Sturtevant's students at California Institute of Technology, began a thorough study of the *bithorax*-like mutations during the late 1940s; his publications began appearing in 1951. However, it is in a later paper (Lewis, 1964) that he clearly outlines genetic methods for the analysis of development; this outline closely coincides with the subsequent chapters of the present text:

1. The earliest method consisted of studying the effect of substituting one allele for another.

2. Utilizing duplications and deficiencies for parts of chromosomes, it became possible to study the effect of 1, 2, 3, and possibly

[*]Subsequent to the preparation of this manuscript, in which frequent reference is made to his research, E. B. Lewis was awarded the 1995 Nobel Prize in Physiology or Medicine—a prize that he shared with two other developmental geneticists, E. F. Weishaus and Christiane Nüsslein-Volhard.

more copies of a mutant allele. In the case of *apricot* and *eosin* (both eye-color mutants) and *bobbed* and *scute* (bristle mutants), extra doses of the mutant allele make their carriers more nearly normal. Muller (1932) referred to such mutants as *hypomorphs:* the mutant allele performs the same function as the wild-type one but does so less efficiently.

3. By utilizing events such as somatic crossing-over, somatic mutation, or the somatic loss of chromosomes (that is, genetic changes that occur in body cells rather than in the more familiar germ line cells), one can obtain mosaic individuals possessing patches of tissue that exhibit contrasting phenotypes; these allow direct comparisons of genetically different tissues and organs in an otherwise identical genetic background.

4. Position effect, the dependence of a gene's expression on alleles at nearby gene loci, provides a powerful tool for the analysis of gene action during development.

5. The last of the genetic methods listed by Lewis is the use of *suppressor mutations:* that is, mutant alleles of normal genes at various loci that lead to normal, rather than abnormal, development in individuals carrying mutant alleles at other gene loci. These genes, of which *su(Hw)*—suppressor of *Hairy wing*—was the first to be discovered, may suppress the mutant phenotype of a single mutant gene or suppress those of many. Lewis (1949) discovered a second suppressor of *Hairy wing* [*su(Hw)*^{2}] that has proved to suppress the phenotypes of mutants at the following loci: *Hw* (hairy wing), *sc* (missing bristles), *dm* (diminutive bristles), *ct* (cut and scalloped wings), *lz* (small eye with fused or irregular facets), *Bx* (long narrow wings), *bxd* (thoracic abnormalities), *ci* (interrupted wing veins), *B* (narrow eye), *f* (irregularly shaped bristles), and *y* (yellow body color). In many instances, *su(Hw)*^{2} suppresses one or two mutant alleles at a given locus but not all—thus showing that the genetic lesions that are responsible for otherwise similar mutant phenotypes must differ. Molecular studies have revealed that *su(Hw)* suppresses mutations that result from the insertion of one type of "jumping" gene ("gypsy") at any of many gene loci.

Returning to Lewis's first point, *bithorax* flies were examined and compared with their wild-type counterparts from the moment of the mutation's discovery. The first descriptions came, as expected, from the Morgan laboratory (Bridges and Morgan, 1923; Morgan et al.,

1925). The examination moved inside the fly when T. Y. Chen (1929) observed that the dorsal metathoracic disc of mature *bithorax* larvae is 60%–65% larger than that of the wildtype; the difference in size is detectable at 40 hours after hatching at 25°C. Subsequent to Chen's investigations, various persons compared the histological appearances of wild-type and *bithorax* flies. Shatoury (1956), for example, noted that in the secondary mesothorax (i.e., the transformed metathorax) neither the dorsal flight muscle nor the lateral group of fibrillar muscles that are present in wild-type flies are to be found. Shatoury then went on to say (p. 235) that "examination of extreme *bithorax* individuals shows that their third legs have acquired the patterns of bristles and hairs characteristic of normal second legs." This changing pattern of normal development was pursued relentlessly by Lewis; his success in unraveling the phenotypic changes associated with *bithorax*-like mutations attests to the value of Barbara Mc-Clintock's oft-cited admonition: "Know your organism!"

Fine-structure genetic analyses

Observation and description are valuable first steps in understanding gene action, but they provide limited information—even when the observations and descriptions are biochemical in nature. Lewis, having accumulated many spontaneous and X-ray–induced mutations at the *bithorax* "locus," conducted a detailed analysis of their map (and chromosomal) positions. His early results are shown in Figure 4-3. Every mutation could be located at one or the other of five loci (designated a, B, C, d, and e), each of which is about 0.01 map unit from the other (one recombinant among 10,000 gametes). The entire complex (which is assigned "crudely" to map position 58.8) is located between two flanking "marker" genes: *ss* (*spineless*) and *Mc* (*Microcephalus*)—somewhat closer to the latter. This complex of three closely linked genes has been traced to their locations on the giant salivary chromosome; these locations are identified in Figure 4-3. Parenthetically, it should be mentioned that many complex gene loci that prove to be composed of subunits (a–e, in the case of *bithorax*) are often found to reside in what can be interpreted as duplicated chromosome bands.

Cells arise only from preexisting cells. Nuclei arise only from preexisting nuclei. Chromosomes arise only from preexisting chromo-

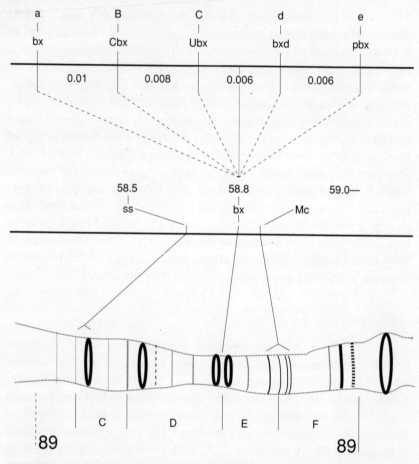

Figure 4-3. The genetic map, giving the relative positions of the postulated five genes at the *bithorax* locus, and the cytological map, showing that this locus falls within a pair of doublet bands that lie between *spineless* (*ss*) and *Microcephalus* (*Mc*). Molecular studies of the DNA of this region have suggested that under one definition of *gene*, there may be *three* genes in this region. (After Lewis, 1964, by permission of Academic Press and E. B. Lewis.)

somes. And genes, as segments of chromosomes, arise only from preexisting genes. Hence, as Bridges inferred in 1918 (see Bridges, 1935), when he discovered duplicated segments in *Drosophila* chromosomes, new genes are normally acquired during an organism's evolution only by the duplication of preexisting genes, followed by a divergence in the functions of the original and duplicated copies. Because the duplication of a chromosomal segment ("gene") generally involves a local error in DNA replication, duplicated genes are often adjacent. Closely linked genes having similar phenotypic effects ("pseudoalleles"; see McClintock, 1944) were used as models by Lewis (1951) in attempting to understand the basis for *bithorax*-like phenotypes. (Today, errors in copying DNA are known to cause numerous small duplications and deletions of DNA at the molecular level as well as of whole genes.)

Position effect

The approach Lewis took in 1951 was the fourth in the outline of genetic methods presented earlier: *position effect*. The classic example of position effect is the *Bar* eye mutation in *D. melanogaster*. The *Bar* (small eye) phenotype is caused by the tandem duplication of a small region of the X chromosome. A normal-eyed female carries two copies of the region, one on each X chromosome. A homozygous *Bar*-eyed female carries four copies, two on each chromosome. Because the duplication is tandem (*abcdabcd*, as opposed to a reverse duplication, *abcddcba*) mispairing of homologous elements followed by recombination in homozygous *Bar* females can result in either reversion to a normal X chromosome or an X chromosome that carries three copies of the duplicated segment (see p. 36). The latter chromosome causes an extreme (*Ultra*) *Bar* phenotype. Females carrying a normal and an *Ultra Bar* chromosome have four copies of the duplicated segment, the same number as homozygous *Bar* females. Nevertheless, the eyes of *Ultra Bar* heterozygotes are consistently smaller than those of homozygous *Bar* females. The effect of the duplicated segment, that is, depends upon its neighbors—upon its position in the chromosome. Hence, the term *position effect*.

Figure 4-4 illustrates (A) the normal wing of *D. melanogaster* and (D) the fly's normal haltere. Figure 4-4B illustrates the haltere of an extreme *bithorax* mutant; Figure 4-4C illustrates the haltere of an extreme *bithoraxoid* mutant. Close inspection reveals that the winglike aspect of the *bithorax* haltere is restricted to its anterior portion; the posterior portion is essentially normal for a haltere. In contrast, the *bithoraxoid* haltere is winglike in its posterior half, but haltere-

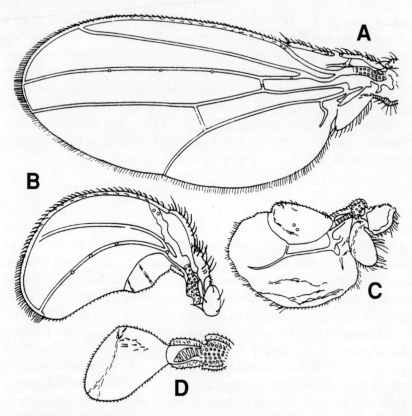

Figure 4-4. Drawings of the normal wing (A) and haltere (D) of *Drosophila melano-gaster;* the former is mesothoracic in origin and the latter, metathoracic. In a *bithorax* mutant, the anterior portion of the haltere becomes winglike in structure (B). In a *bithoraxoid* mutant, the posterior portion of the haltere becomes winglike (C). Note that the winglike alterations in B and C correspond to the anterior and posterior portions of the normal wing, respectively. (After Lewis, 1951, by permission of Cold Spring Harbor Laboratory and E. B. Lewis.)

like in its anterior portion. By knowing his organism, Lewis was able to see that the fly's thoracic segments have distinguishable anterior and posterior portions.

Lewis's study of the morphological effects of *bithorax* mutants was extended to flies that were identical in genotype except for the ways in which alleles were associated. Figure 4-5 shows the halteres of flies of six different genotypes. The genotypes of the upper and lower members of each of the three pairs differ only in the patterns of asso-

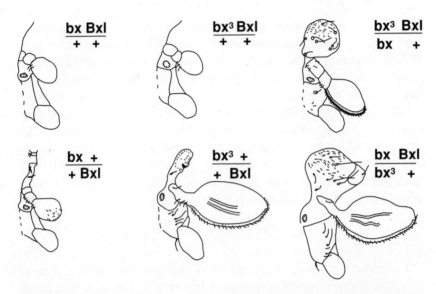

Figure 4-5. Diagrams of the halteres and nearby structures of flies with the indicated genotypes. Note that, in each vertical pair, the genotypes are identical except for the position or distribution of the mutant alleles. *bx, bithorax; Bxl, bithorax*-like. The normal allele in each instance is indicated by +. Differences in the phenotypes exhibited by members of the vertical pairs are ascribed to *position effects*. (After Lewis, 1951, by permission of Cold Spring Harbor Laboratory and E. B. Lewis.)

ciation of the alleles. The halteres of flies carrying a wild-type (++) chromosome are essentially normal; *bithorax*-like (*Bxl*) results only in the development of one or two otherwise unexpected bristles on the anterior edge of the haltere. On the contrary, in combination with a chromosome that bears *bithorax* (*bx*) or *bithorax-three* (*bx³*), *Bxl* has a more pronounced effect—especially when the homologous chromosome carries *bx³*. In this case, the posterior portion of the haltere is greatly enlarged and winglike. The rightmost pair of drawings suggest that *Bxl* has a much greater effect on the abnormal development of the posterior portion of the metathorax when *bx³* is on the homologous chromosome than when it is on the same chromosome as *Bxl*. One of the extreme observations (not illustrated) concerns the *bxd/bxd* homozygote or *Bxl+/+bxd* heterozygote: "In such types, the posterior portion of the metathorax comes to resemble the posterior portion of the mesothorax, the anterior metathorax remaining unchanged. . . . At the same time there is always a thoracic-like modification of the first abdominal segment, one of the most striking results of which is the occasional production of a pair of first abdominal legs, which arise in addition to the normal three pairs of thoracic legs" (Lewis, 1951, p. 165). Thus, as Lewis concluded, the posterior half of each of the fly's segments appears to be more closely related embryologically to the anterior portion of the segment that follows than to the anterior portion of its own segment. The means by which he arrived at this conclusion involved the assumption that the complex of five loci that constitute the *bithorax* "locus" have arisen by duplication as suggested by Bridges; following their origins, however, their functions have diverged.

Recapitulation

The picture that has emerged from Lewis's many and careful morphological observations is shown in Figure 4-6. The wild-type fly is illustrated by a simple diagram at the top. It consists of a head, thorax, and abdomen. The three thoracic segments (pro-, meso-, and metathorax) bear legs. The last two thoracic segments bear wings (*mesothorax*) or halteres (*metathorax*). The abdominal segments bear no appendages.

The morphological effects of mutations at the five genes at the *bithorax* "locus" are illustrated in the lower portion of Figure 4-6. *Bithorax,* itself, reduces the posterior portion of the haltere (a). *Contrabithorax* (B) removes the posterior portion of the wing as well as both the anterior and posterior portions of the haltere. *Ultrabithorax* (C) is lethal when homozygous, so its potential effect on morphology is unknown. *Bithoraxoid* (d) reduces the anterior portion of the normal haltere while expanding the posterior portion, but, by inducing (or allowing) the first abdominal segment to differentiate into a thoracic one, it causes a second (posteriorly) enlarged haltere to develop. *Bithoraxoid* also results in the development of a leg on the transformed abdominal segment. *Postbithorax* (e) allows the enlargement of the posterior portion of the haltere and also the transformation of the first abdominal segment into an appendageless thoracic segment. The choice of verbs *removes, reduces, induces,* or *allows* is not a trivial matter. Subsequently, we shall see reasons for believing that the normal gene products of these genes *suppress* normal abdominal segmentation, thus *allowing* thoracic segments to develop.

WILD-TYPE BODY SEGMENTATION

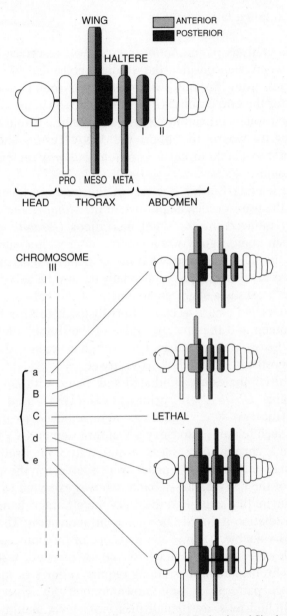

Figure 4-6. Diagrammatic representations of a normal fly (top) and flies homozygous for mutations at each of the five (or, currently, three) genes that constitute the *bithorax* locus (see Figure 4-3). Note that, in order to represent the mutant phenotypes, one must represent each body segment as consisting of an anterior and a posterior portion; this need reappears in discussing embryonic *parasegments* (Figure 4-11). Winglike modifications of the haltere are shown as elongations in these diagrams. (After Lewis, 1963, by permission of *American Zoologist* and E. B. Lewis.)

Transvection à la Lewis

Lewis's analysis of the *bithorax* locus will resurface later in the chapter when we consider the studies molecular biologists have carried out, using his observations as their foundation. Even the routine (for the era) practice of locating genes on giant salivary gland cell chromosomes proved to be essential because it told the molecular biologists where to obtain the DNA (using chemical procedures not available to Lewis) from the *bithorax* region of the fly's chromosome.

Before leaving Lewis's analysis of the *bithorax* locus, we might mention the use to which he put two of the *bithorax* alleles in detecting X-ray–induced chromosomal aberrations. *Ultrabithorax* (*Ubx*) is lethal when homozygous; when heterozygous with a wild-type allele (*Ubx*/+), the haltere is somewhat larger than normal and possesses one or two bristles that do not normally occur on a haltere. A second allele, bx^{34e}, causes a slight enlargement of the halteres in homozygous carriers and a narrow band of bristly tissue to form between the fly's abdomen and the posterior edge of the thorax. Flies heterozygous for the two mutant alleles, Ubx/bx^{34e}, have enlarged halteres but lack the postthoracic band of bristly tissue.

If bx^{34e}/bx^{34e} males are irradiated and mated with *Ubx*/+ females, one obtains Ubx/bx^{34e} offspring. Lewis (1954) used attached-X females (the two X chromosomes were attached to a single centromere; such females also carry a Y chromosome: XXY) so that the irradiated X chromosome was transmitted from the irradiated father to his sons; only Ubx/bx^{34e} males were scored for the presence or absence of the hairy band—a band that was expected to be absent.

Every male that exhibited the hairy metathoracic band proved to carry a radiation-induced chromosomal aberration. This screening method is valuable, because the presence of the induced rearrangement is detected in an individual fly; the usual genetic tests for newly induced chromosomal aberrations require *cultures* of flies in which the linkage patterns of well-known mutant genes are drastically altered. To find, let's say, 10 male flies with a hairy metathoracic band out of 1000 males examined requires much less time and effort than setting up 1000 cultures and then examining the progeny in each vial, scoring linkage relationships.

Although it was true that all Ubx/bx^{34e} flies that exhibited the hairy

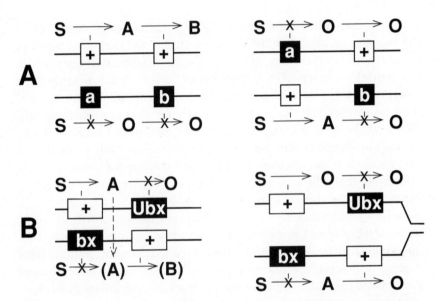

Figure 4-7. Schematic interpretation of an empirical observation: the phenotypes of +*Ubx*/*bx*+ flies differ depending upon whether somatic pairing occurs at that chromosomal region (B, left) or is prevented by a chromosomal rearrangement (B, right). In Figure 4-7A, a substrate (S) is assumed to be converted into substance A by the wild-type allele of *a;* substance A is then converted into substance B by the wild-type allele of *b*. These events are assumed to occur with no diffusion between homologous chromosomes; hence, the *trans-* arrangement (A, right) of the mutant alleles prevents the formation of substance B. To account for the different phenotypes observed at the left and right in Figure 4-7B, one must assume that substance A can diffuse from one homologue to the other if they are somatically paired; otherwise, no diffusion is possible. (Adapted by permission of the University of Chicago Press and E. B. Lewis from Lewis, 1954, *American Naturalist* 88:225–239.)

thoracic band carried chromosomal rearrangements, not all chromosomal rearrangements were to be found only in such flies. Males without the hairy band also proved to carry radiation-induced aberrations. Upon examining the locations of the chromosomal breakage points, Lewis made a startling discovery: every aberration that resulted in the hairy band was one that interfered with the normal pairing of the homologous regions at the site (58.8, chromosome 3) of the *bithorax* locus.

Figure 4-7 illustrates Lewis's explanation for his observation. At the top (Figure 4-7A) is illustrated the general difference between the *cis-* (same strand) and *trans-* (opposite strands) arrangement of two mutant pseudoalleles (such as those at the *bithorax* locus). When the two mutant alleles are on the same strand and their wild-type (+) counterparts on the other, the wild-type strand functions normally: S (substrate) is converted to A (product of the left allele), which is then converted to B (the end product) by the second allele. In contrast, the strand carrying the two mutant alleles produces nothing (0). However, because it forms B, the "wild-type" strand exhibits dominance, and development (or metabolism) is essentially normal. The two mutants, when in the *trans-* arrangement, have a quite different effect: Neither chromosome can produce substance B; therefore, development is abnormal—perhaps lethally so.

Before proceeding to the lower diagram in Figure 4-7, we must make a point about dipteran cytology: chromosomes of *Drosophila* and many other flies exhibit what is called somatic pairing. This pairing is especially pronounced in the salivary gland cells of most *Drosophila* species. The pairing of the homologous salivary cell chromosomes is so tight that the observer generally forgets that each giant chromosome comprises two homologues. Only when the homologues exchange partners (as they must in the pairing within inversion or translocation heterozygotes) is one reminded of the paired structure of these salivary gland chromosomes. However, even in somatic cells such as oogonia or neural cells, metaphase configurations reveal the homologous chromosomes lying side by side: two X chromosomes (female), two pairs of V-shaped autosomes, and a pair of dots in the case of *D. melanogaster*.

The X-ray–induced aberrations that went undetected by Lewis's screening procedure were those that allowed the *bithorax* region of the third chromosome to pair normally (Figure 4-7B, left). Those that resulted in the hairy metathoracic band were those whose breakage points interfered with that normal pairing (Figure 4-7B, right). Thus, because *Ubx* and *bx*^{34e} were in the *trans-* configuration (compare with Figure 4-7A, right), Lewis had to assume that a gene product (A) made by the wild-type allele of *bx*^{34e} diffuses to the other chromosome, where the normal allele of *Ubx* transforms it into B, the normal end product. In contrast, when the somatic pairing of the two homologues is prevented, substance A is unable to reach the homologous strand; thus, neither chromosome can produce the necessary end product B. Lewis called this phenomenon, of which several other cases are now known in *Drosophila,* "transvection." The explanation is not yet known. Perhaps large molecules—polymerase molecules or segments of mRNA—that can diffuse only short distances are involved, or, if one wishes to speculate, perhaps somatic pairing of chromosomes leads to an intimate intermingling of DNA strands, some of which through faulty replication reconstitute normal (*cis-*) strands at the *bithorax* locus.

A pause in the narrative concerning mutants at the *bithorax* locus is now called for. The studies carried out by E. B. Lewis over a period of 30–40 years are among the most thoughtful and most extensive in what may be called "classical genetics." That these studies are rooted in a study of morphological differences that correspond in every way to the round and wrinkled or dwarf and tall peas of Mendel has by now been forgotten. The genetic methods recommended and used by E. B. Lewis were methods that were developed by early *Drosophila* geneticists for just this purpose: to manipulate genetically different strains of flies in ways not intended by nature. The genetic system of the fly as well as its biological characteristics (ease of handling; many progeny per female) were the starting points. Attached-X chromosomes, those sharing a single centromere, either were man-made or were natural accidents that were discovered and saved by shrewd experimentalists. The mutagenic effects of X rays and gamma rays were exploited in pressing the investigation into the action of the various alleles at the *bithorax* locus.

Before moving on to a consideration of procedures of which Lewis could only dream during the time he carried out his studies, we might ask, In what other organism did investigators imitate (or duplicate) investigations of the sort performed by Lewis using *D. melanogaster?* Until recently, the answer, for all practical purposes, has been, None. S. G. Stephens (1946, 1950), in studying in species of cotton complementary recessive alleles whose only observable action is to kill interspecific hybrids when homozygous (homozygosity within any one species produces no discernible effect), concluded that the responsible locus was compound—that is, it arose as a gene duplication. Thus, he (Stephens, 1951) (and the corn geneticists working with the R- and A-loci of *Zea mays*) could join E. B. Lewis in discussing "pseudoallelism"—a phenomenon associated with duplicated gene loci, a phenomenon related to position effect (see Figure 4-5).

Molecular aspects of development

The construction of an egg by a female organism is frequently as time consuming as the development of an individual following fertilization. A female toad may spend the better part of a year developing eggs; tadpoles emerge only days after a toad's eggs are fertilized. Female human beings have their entire allotment of oocytes at birth; oocyte maturation (which includes the development of follicular cells) may take 12 or more years; gestation time is nine months. In *Drosophila,* the earliest mature oocytes appear during the pupal stage just before emergence of the adult female; attaining sexual maturity prior to mating requires about 24 hours; the first-instar larva hatches from the egg 24 hours after fertilization.

The role of the female in the organization of her eggs' cytoplasm is revealed by a number of inherited characteristics that are lumped under the rubric "maternal effects." The left-handed or right-handed coiling of a snail's shell is determined not by the affected individual's genotype but by that of its mother. It is the mother's genotype (by virtue of the pattern with which proteins are laid down in the egg) that determines whether the first two divisions of the fertilized egg initiate a left- or a right-handed whorl (Figure 4-8). *Grandchildless* (*gs*) mutants are known both in *Drosophila melanogaster* and *D. subobscura.* Although the *grandchildless* females are capable of bearing progeny, they are incapable of producing eggs that develop into *fertile* offspring—hence, no grandchildren. The missing element in eggs produced by *gs/gs* females are the densely staining "polar granules" that occur in normal eggs and that, during embryonic development, define the area from which germinal cytoplasm is derived.

A number of molecular techniques have been developed that permit one to examine the development of *Drosophila* eggs and larvae in a manner unimaginable prior to the 1970s or even 1980s. One of these involves nurse and follicle cells (Figure 4-9). These are cells of maternal origin that surround the egg and, in assisting in its maturation, transfer their own RNA as well as nutrient substances into the egg cytoplasm. In situ hybridization using radioactive RNA or DNA probes reveals that, during the maturation of the egg in the ovary, follicle cells transfer RNA into the egg—one type (*bicoid*) at the anterior end, another (*nanos*) at the posterior pole. Mutant females lacking the wild-type *bicoid* allele produce eggs that develop

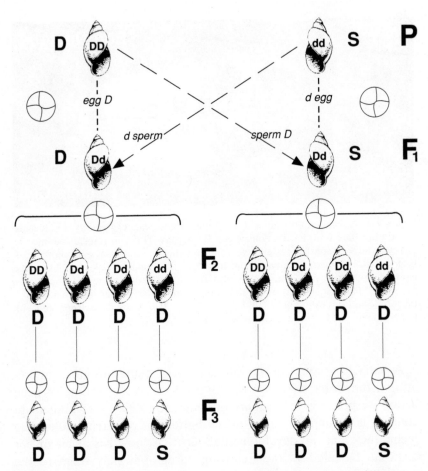

Figure 4-8. An illustration of the maternal effect, which determines the direction of coiling in the shell of *Limnaea peregra*. Females homozygous (*DD*) or heterozygous (*Dd*) construct eggs (small, four-parted circles) whose early divisions result in a right-hand (dextral) coil; *dd* females, in contrast, construct their eggs so that early divisions lead to a left-hand (sinistral) coil. Note that this maternal effect delays the appearance of the expected 3:1 Mendelian ratio to the F_3 generation, rather than revealing this ratio in the F_2 generation.

into embryos lacking heads and thoraxes; females lacking the wild-type *nanos* allele produce eggs that develop into embryos possessing heads and thoraxes but lacking abdomens. Thus, the anterior-posterior axis of the egg is determined by maternal genes acting by way of ovarian (follicle) cells and substances (mRNAs) that they deposit in the egg.

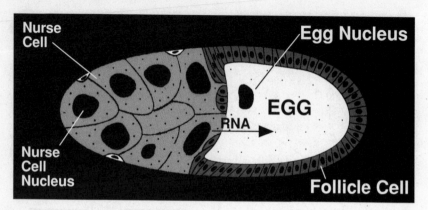

Figure 4-9. A diagram illustrating a developing *Drosophila* egg and its closely associated nurse and follicle cells. Follicle cell RNA (*bicoid*) that is transferred into the egg cytoplasm establishes the embryo's anterior (head and thorax) end; another follicle cell RNA (*nanos*) similarly transferred at the opposite pole establishes the embryo's posterior end. Nurse cells also transfer RNA into the developing egg (arrow). Thus, these aspects of the developing embryo are under maternal control. (After Slack, 1991. Copyright © 1991 by Cambridge University Press. Reprinted with the permission of Cambridge University Press.)

A detailed account of the interactions of numerous genes in creating a functional fly is not required for an understanding of gene action—especially of the *study* of gene action. Of more immediate concern is an understanding of the battery of research (molecular) techniques that the developmental geneticist now has at his or her disposal. Detailed as the following paragraphs may seem to the reader, they cannot do justice to the matter of *development* in higher organisms. An appreciation of the latter can be obtained from Peter Lawrence's small book *The Making of a Fly,* or, for a readable overview of developmental research on many organisms (including that initiated by E. B. Lewis), from the Howard Hughes Medical Institute publication *From Egg to Adult.*

From E. B. Lewis's exceptionally fine morphological and genetic analysis of the *bithorax* mutants, molecular geneticists obtained the clues that permitted them to zero in on the chromosomal region that is involved in the proper development of the metathorax and its neighboring segments. A second chromosomal region (see Figure 4-10) contains the locus, *Antennapedia,* that exerts a corresponding control over the anterior segments of the fly's body. Knowing where

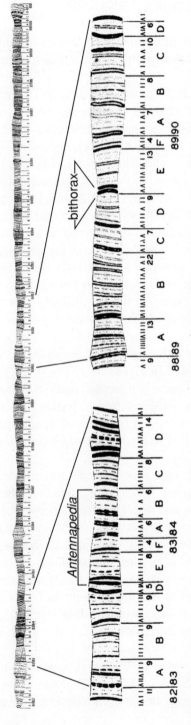

Figure 4-10. Salivary chromosome maps illustrating the wealth of cytological detail that these giant chromosomes reveal. At the top is illustrated the right arm of chromosome 3; this represents about 20% of the entire genome of *Drosophila melanogaster*. The two inserts illustrate the detail that is visible at the *Antennapedia* and *bithorax* loci, two homeotic loci—i.e., loci at which mutations tend to convert normal appendages into appendages usually found on different body segments. The *bithorax* locus was also shown in Figure 4-3. Chromosomal rearrangements that alter the relative positions of the *Antennapedia* and *bithorax* loci within the fly's genome do not necessarily alter their functions during development.

to look for these gene loci, molecular geneticists were able to isolate and clone the DNA that corresponds to these two loci, and to neighboring ones as well.

With the cloned DNA in hand, one can then perform a sequence analysis and, using the codon for methionine (AUG) as a guide, search for one or more open reading frame(s)—that is, lengthy sequences of nucleotides within which each successive set of three (codons) specifies an amino acid. This DNA can be put to two uses: first, it can be used as a probe to locate its corresponding mRNA; second, it can be used to synthesize mRNA and, in turn, the corresponding protein. Antibodies to this protein can be obtained by injecting the protein into a rabbit. Finally, one might note that fluorescent antibodies to immunoglobulins (antibodies) are available; hence, one can locate where within a section of tissue a given protein is located: the investigator exposes tissue to an antibody for a given protein, washes off the excess antibody, and exposes the washed tissue to the fluorescent antibody that attaches to the already-bound antibody.

The sorts of uses to which these procedures can be put in the study of gene action are illustrated in Figure 4-11. The three genes whose roles are followed in this figure are *bicoid* (*bcd*), *Krüppel* (*Kr*), and *hairy* (*h*). Radioactive probes were used to locate the mRNA transcripts for wild-type alleles at each of the three genes; the three sections in each horizontal row are adjacent ones cut from the same wild-type embryo.

In the case of *bicoid,* one sees that ample mRNA has been deposited in the anterior of the egg by follicle cells within the mother's ovary. With time, this mRNA diffuses through the anterior portion of the egg and gradually disappears. Studies of the protein synthesized from the *bicoid* mRNA show that its synthesis and distribution follow the spread of the mRNA itself.

The mRNA of the *Krüppel* (German for "cripple," so named because the mutant alleles at this locus result in flies with thoracic abnormalities, often with missing wings or legs) begins to appear in the midline of the embryo; it persists through the three hours or so covered by Figure 4-11.

The *hairy* gene (*h*) produces mRNA that appears only after the last of the early nuclear divisions and as the nuclei (now located at the outer surface of the egg) become enclosed by cell membranes.

minutes

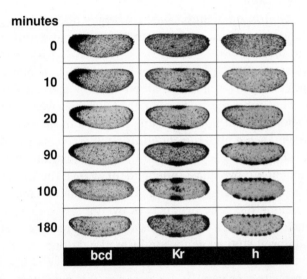

| | bcd | Kr | h |

Figure 4-11. Diagrams based on photographs of *Drosophila melanogaster* eggs that, at regular intervals after being laid, were exposed to radioactive probes in order to reveal the mRNA transcripts of the wild-type alleles at four loci: *bicoid* (*bcd*), *Krüppel* (*Kr*), *hairy* (*h*), and (not illustrated) *wingless* (*wg*). Each row shows three adjacent sections cut from the same embryo; the six sections in vertical columns are from successively older embryos. The *bicoid* transcript is found initially at the anterior portion of the embryo, where it has been deposited by maternal follicle cells; this material diffuses through the anterior portion of the developing embryo, being transcribed into a bcd protein during this time, and disappears. The *Krüppel* transcript is concentrated at the midline of the embryo, where it interacts with products of the *hairy* locus; the latter form seven ventral patches (parasegments). The transcript of the *wingless* locus, as described in the text, appears late and forms fourteen stripes that correspond to the embryonic parasegments. Each embryonic parasegment gives rise to the posterior half of one larval or adult body segment and the anterior half of the next one. Only seven alternating parasegments persist to give rise to body segments. (Adapted with permission of Macmillan Magazines Limited and P. W. Ingham from Ingham, 1988, *Nature* 335:25–34. Copyright 1988 Macmillan Magazines Limited.)

This mRNA is distributed in clearly defined embryonic segments (parasegments). "Parasegment" is emphasized because each parasegment gives rise to the posterior half of one body segment and the anterior half of the next one. (This relationship between embryonic and postembryonic segments had been earlier deduced by Lewis from the phenotypes of his mutant flies.)

Finally, the wild-type allele of *wingless* (not illustrated in Figure 4-11) produces mRNA that is found in fourteen stripes that indicate the early organization of embryonic parasegments. (Only every other one of the fourteen goes on for further development, thus reducing fourteen to seven.)

The parasegments that are seen in Figure 4-11 (*hairy*) give rise to the subsequent body segments, albeit not on a one-to-one basis. However, what form the body segment takes and what appendages it will develop depend upon the gene products of the two homeotic genes, *Antennapedia* and *Ultrabithorax,* that were discussed earlier. The establishment of a segment and the determination of the *nature* of that segment are two quasi-independent events. The whole cascade of events is far from being understood; the stimulatory and inhibitory reactions among the gene products of even a dozen or so gene loci stagger one's mind. Such interactions are not limited to closely linked genes. Many well-studied genes are located near one another on the fly's third chromosome (map distances 46 to 48.4). Whether this distribution reflects a propensity for genes in this neighborhood to be involved in developmental and morphological matters or whether it reflects a bias on the part of molecular geneticists to concentrate their efforts on a well-studied region of a chromosome is not clear. What is clear, and has been clear to most geneticists of the twentieth century, is that, although a mutation at one gene locus can cause a specific abnormality, the wild-type allele at that locus does not cause the absence of that abnormality. A wild-type allele alone does not—indeed, cannot—cause normalcy.

Summary

The purpose of this chapter has been to outline the evolution of procedures used by geneticists in attempting to understand a problem that invites solution. Indeed, the problem itself evolves—it does not remain static. With morphological mutants, the problem evolves from their patterns of inheritance (a problem that is related to the search for the gene) to the material (physiological or biochemical) causes of the abnormal patterns of development. The very earliest morphological abnormalities observed still pose problems for the molecular biologists of today. Indeed, although we have used the *bithorax* locus of *Drosophila melanogaster* as an extensive case study, an entire battery of human disorders could have been cited instead. Human disorders have been recognized as such for millenia; clay figures of deformed persons were created by artists in ancient civilizations. Seeking and understanding the causes of these disorders are goals of current research done in many laboratories worldwide.

5 Color

Early plant hybridizers, even those preceding Mendel, traced the inheritance of morphological traits (shape) and color. Even Darwin, during his analysis of the effects of cross- and self-fertilization in plants, noted the differences in color that sometimes appeared in his material. Whereas many plant hybridizers noted the contrasting colors but failed to count the numbers of individuals in each category, Darwin counted (he saw an excellent 3:1 ratio in one case), but, because he was engrossed in another problem, he failed to see (or to contemplate) a pattern of inheritance. Mendel, of course, was a meticulous, professional hybridizer who not only concentrated on the pattern of inheritance but also realized that numerical data were necessary for revealing the basis of that pattern.

Shape and color. Those were the two aspects of an individual's phenotype that early geneticists could observe and classify with little difficulty. Once patterns of inheritance were established and research interest shifted to gene action, to physiological genetics, traits involving color and shape were the most promising. Contrasting patterns that seem clear-cut at one level of inquiry become hopelessly complex when pursued further (see Figure 5-1). The interactions of alleles R and r (rose comb versus non-rose in chickens) with P and p (pea comb versus non-pea) (Figure 5-2) form a classic example of epistasis (the interaction of alleles at different gene loci) for beginning genetics students (9/16 walnut comb: 3/16 rose comb: 3/16 pea comb: 1/16 single comb in the F_2 generation), but there the matter rests. No one would consider attacking the "How-come" question in this case; far too much time, effort, and expense would be involved for the knowledge gained.

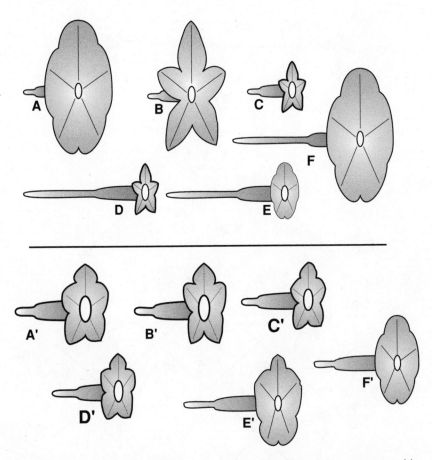

Figure 5-1. Diagram indicating that the extreme recombinants that are expected in the F$_2$ generation of a cross between *Nicotiana langsdorfii* and *N. alata* were not encountered among approximately 500 individuals examined. The flowers of the two species differ greatly in corolla tube length, limb (the expanded portion of the corolla) width, and the degree of lobing. Upper diagrams illustrate the six possible extreme recombinants; lower diagrams illustrate the nearest approach to each of the theoretical extremes (A′ → A, B′ → B, etc.). Note that for each of the three phenotypic traits, even the extreme variants approximate an average value. (After Anderson, 1949.)

Colored compounds: Some chemistry

Color, especially color arising from organic pigments, is a simpler aspect of an organism's phenotype than is its shape. Organic chemists, from 1865 when the structure of benzene was reasonably well known,

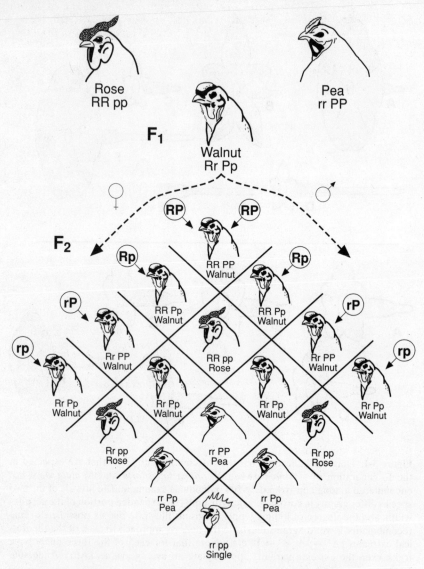

Figure 5-2. The inheritance of comb form in the chicken: a classic example of gene interaction that has been cited in introductory genetics texts for decades. Although this trait is excellent for illustrating both patterns of inheritance and the epistatic interactions of genes at different loci, it provides little opportunity, if any, for studying the nature of gene action: what are the primary actions of the dominant and recessive alleles at these two loci that lead to the physical shape and texture of the bird's comb?

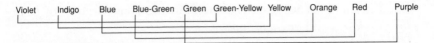

| Violet | Indigo | Blue | Blue-Green | Green | Green-Yellow | Yellow | Orange | Red | Purple |

Figure 5-3. "White" light is a mixture of wavelengths of electromagnetic radiation within the "visible" range in proportions such that they are interpreted as being white. The absorption of any one wavelength by a chemical substance results in that substance's giving the sensation of possessing the complementary color. Thus, an absorption band in the violet region causes a dye to appear green-yellow.

began a systematic search for dyes that could be made from coal-tar products. This search led to a theory of color (see Figure 5-3) as well as to an understanding of the relation between color and the chemical structure of organic molecules. In general, a molecule is "colored" if it possesses an absorption band within the range of visible electromagnetic radiation; the absence of this wavelength converts the original mixture ("white") into a mixture that appears to be colored. Picric acid, for example, absorbs in the indigo range, hence it appears to be yellow (Figure 5-3). Organic molecules that appear to be colored and, therefore, may be useful as dyes, are unsaturated molecules. Common *chromophore* groups are:

$$-N\underset{\displaystyle O}{\overset{\displaystyle O}{\big<}} \qquad -N{=}O \qquad -N{=}N- \atop O$$

$$={\big<}{=}{\big>}= \qquad {\big<}{=}{\big>}= \qquad -N{=}N-$$

Less common, and weaker ones, are:

$$\overset{O}{\underset{\|}{-C-}} \qquad \overset{S}{\underset{\|}{-C-}} \qquad {>}C{=}N-$$

$$-C{=}C-$$

Compounds that exhibit color are not necessarily useful as dyes; the latter property depends upon portions of the molecule (radicals) that can form salts, thus making the compound soluble. Such *auxochromic* radicals are OH, NH_2, NHR, and NR_2, where R stands for a double-ring organic acid. The actual color of an organic compound is determined in complex ways by the number and configuration of both *chromophores* and *auxochromes*.

The above account of colored compounds and dyes was stimulated by the extent to which natural pigments have been used by persons both ancient and modern. Indigo, once the most important commercial dye, was originally extracted from small plants (*Indigofera* spp.) as early as 1500 B.C. By the end of the nineteenth century, natural indigo had been replaced commercially by the synthetic counterpart. A relative of indigo, royal purple, is recovered naturally from molluscs and other shellfish: 12,000 g of molluscs may yield 1.5 g of dye. Chemically, royal purple is merely a brominated indigo. Organic dyes and other colored organic compounds have attracted the attention not only of geneticists but also of chemists and industrialists. A great deal of genetic research on pigments is supported, of course, by those who have a financial interest in particular research materials—materials that make plants or animals especially valuable or that can be extracted for commercial purposes.

Biological basis for pigmentation

Pigments are useful not only for human beings. They have a role in the organisms that synthesize them. Pigmented microorganisms are, as a result of pigmentation, somewhat protected from otherwise dangerous (inactivating or mutagenic) ultraviolet radiation. Plants have developed variously colored flowers in attracting (within a highly competitive world) efficient insect and bird pollinators. Nor should one ignore the green of chlorophyll, the means by which solar energy is trapped for use by all living things. Animals use colors both for attracting attention (sexual displays, a form of communication) and for concealment (camouflage).

These introductory comments can be conveniently terminated by stressing that the *action* of genes that determine either the shape or the color of organisms is presumably identical. That is, the gene acts by specifying (via mRNA) a particular protein. The transcription of the gene's DNA occurs at the proper time and within the proper cells in response to signals arriving from elsewhere: from other genes, from extracellular hormones, or even from the external world in the form of daylight or ambient temperature.

The advantage color has over shape as a means for studying gene action lies in the (usual) chemical nature of the pigment. The parenthetic qualification is needed because some colors (the blue of blue jays and the spectacular display of peacocks, for example) are physical colors that arise, like the color of oil films or of opals, from the refraction of light from and through films of molecular dimensions. Morphological structure depends upon the physical orientation of complex protein molecules. The chemical nature of pigments depends upon chromophores and (perhaps) auxochromes. Differences between related pigments are differences that may arise through a variety of enzymatic steps. The presence or absence of an enzyme is a genetic alternative. Some pigments are pH indicators; thus, the color of an organism may be a function of pH, a phenotypic characteristic most likely determined by the genetic milieu of the organism. On the other hand, to the extent that an enzyme can or cannot carry out its function at a given pH, this ability may depend upon the amino acid composition of the enzyme—a dependence upon the base composition of DNA.

An overview of what follows

Public speakers are advised to warn their audiences what they intend to say, to say it, and then to review what they have said. The spoken word is ephemeral: once uttered it is gone. The written word persists and can be reviewed at will. Nevertheless, a preview of the coming sections may prove helpful; even a brief terminal summary may also serve a useful purpose.

The early work on color preceded a subsequent era characterized as *biochemical genetics*. Information was obtained by the method known as "eyeballing." One made careful observations, first, as exemplified by Sewall Wright's analysis of the coat colors of guinea pigs, by the examination of colors resulting from different gene combinations and, second, from experimental manipulations, as exemplified by Beadle and Ephrussi's analysis of eye color in *Drosophila melanogaster*. Flower color in plants has served as a genetic tool from the outset; although he did not choose flower color for study, Mendel did mention "the position, color, and size of the flowers" as constituting differences among various forms of peas. Further, he mentioned the absolute correlation that exists between seed-coat and flower color: white seed-coat with white flowers and colored (gray or brown) seed-coat with violet-purple flowers. Chemically, pigments represent distinct compounds; even ephemeral pigments that serve as pH indicators do so because they gain or lose hydrogen ions. The study of flower pigments, however, tends to become an exercise in the chemistry of the pigments, themselves, rather than in genetics. J. B. S. Haldane (1942), in a book titled *New Paths in Genetics,* devoted many pages to plant pigments and their interrelations; organic chemists provided botanists with many excellent research tools. The primitive state of genetic—as opposed to chemical—knowledge is revealed by the following statement from Haldane (1942, p. 60): "It is not unreasonable to expect that enzymes will be found among the immediate products of gene action. [One may speculate] that the process by which genes produce their immediate products is the same as that by which they reproduce themselves, and that the antigen produced by a gene ... differs from the parent gene essentially in not being anchored to a chromosome."

Additional insight into gene action was provided by pigmentation *patterns;* these are patterns formed by enzymatically synthesized pigments—not the green and white patterns caused by somatically segregating normal and defective chloroplasts. Studies of patterns may be traced from early work on somatic mutations in a plant such as *Delphinium,* through studies on seed and plant color in maize, and finally to the mottled kernels that prompted McClintock's research on mobile genetic elements ("jumping genes"). These elements, through the utilization of molecular techniques, have led to an understanding of the exquisite underpinnings for the action of many genes.

Dominance

Of the seven genes that were studied by Mendel, one allele at each locus concealed the phenotype that was characteristic of its alternative allele. Thus, round seed form was dominant to wrinkled, yellow albumen was dominant to green, inflated pod was dominant to constricted, and tall stems were dominant to short ones. The consistency with which this pattern occurred in Mendel's experiments was so impressive that I. I. Schmalhausen, in one of several pre-1900 references to Mendel's studies, emphasized *dominance*, not "Mendelian" ratios, as the chief of Mendel's discoveries.

Flower color provided perhaps the clearest example of *incomplete* dominance; a diluted pigment is more easily understood than an intermediate shape. Correns, one of Mendel's rediscoverers, encountered a clear example of what was once—and misleadingly—called "blending inheritance." Working with *Mirabilis jalapa* (the four-o'clock), Correns crossed individuals bearing red flowers with others bearing white ones, both of these varieties having been shown earlier to be true breeding. The resultant hybrid (Figure 5-4) bore pink, rather than red or white, flowers. When the pink-flowered hybrids were crossed, their offspring bore red, pink, and white flowers. The pinks were twice as numerous as the reds and whites; the latter types were equally numerous. Hence, the F_2 generation revealed the traditional 1:2:1 Mendelian ratio. The intermediate color of the heterozygous individuals allowed them to be identified as such in both the F_1 and F_2 (or later) generations. Thus, almost simultaneously with the rediscovery of Mendel's paper, the dominance that he had encountered in every trait was shown not to be an essential feature of gene action (see Figure 5-5). Indeed, close examination of heterozygous individuals that seemingly exhibit a "dominant" trait (as distinct from hybrid vigor) often reveals that the trait is less extreme than it is in the homozygous dominants. *Dominance*, that is, is frequently a term of convenience. Muller (1950) recounted such subtle differences while arguing that *all* "recessive" mutations with deleterious effects on the health of human beings are harmful as well to their heterozygous carriers. His examples (chosen of necessity from an experimental organism) included the smaller amount of red pigment in flies heterozygous for the white allele than in homozygous wild-type flies and the commonly observed 3%–5% decrease in survival of flies carrying "recessive" lethal genes.

Pigmentation and pigmentation patterns in mammals

The fur color of wild mammals is controlled by a constellation of genes, some of which had been recognized as being Mendelizing factors as early as 1901. These genes, because in concert they affect a single aspect of an animal's phenotype—coat color—are seemingly

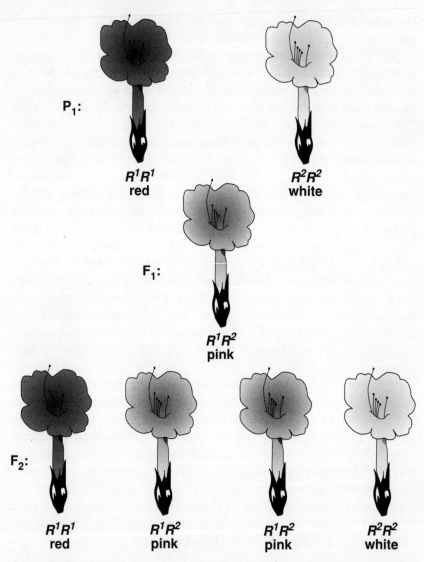

Figure 5-4. The inheritance of flower color in four-o'clock (*Mirabilis jalapa*). Unlike the traits chosen for study by Gregor Mendel, in each of which one of the two contrasting states was dominant, the heterozygous, R_1R_2, four-o'clocks are pink. The term *blending inheritance* that is often applied to this and other examples of *incomplete dominance* is misleading; the alleles that are passed on in the gametes of heterozygous individuals are in no way affected by the intermediate flower color of these individuals.

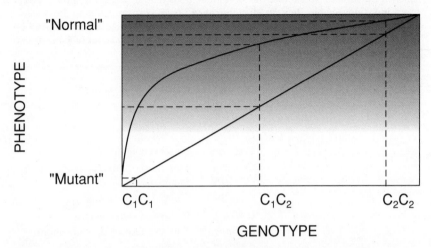

Figure 5-5. A generalized scheme illustrating possible relationships between genotype (C_1C_1, C_1C_2, C_2C_2), enzyme concentration, and phenotype (e.g., amount of pigmentation). Whereas the phenotype may be directly related to the amount of pigment (red, pink, and white), the amount of pigment need not be directly related to the amount of functional enzyme. Even if the amount of the latter were linearly related to the number of functional C_2 alleles (0, 1, and 2), because of feedback and other metabolic controls, the amount of pigment formed in the presence of one C_2 allele may be nearly as great as that produced by two C_2 alleles (dominance, as in *Drosophila* eye color). On the other hand, especially in the absence of feedback controls, the amount of pigment may be proportional to gene dosage (straight line).

more complex than the combinations of distinct pea phenotypes (e.g., tall, inflated pod, and wrinkled seed shape) that confronted Mendel in his studies, but they are not. The alleles at each locus have their contrasting effects, and the effects of one locus constrain the outcomes available to other loci. Such interactions are regarded as *epistatic.*

Table 5-1 lists the phenotypes that are generated by the contrasting dominant and recessive alleles at six gene loci affecting coat color in the mouse. The corresponding genes are found in many mammals and are largely responsible for the coat colors of such animals as black panthers, spotted horses, and variously colored house cats. The six loci can be reviewed separately; following that, only patience is required to generate the information listed in Table 5-1.

The C locus. The dominant allele, *C,* is responsible for the presence of colored pigment; the recessive allele, *c,* results in the absence of

Table 5-1. The coat-color phenotypes that result from the interaction of six coat-color genes in the mouse. Notice that the 32 combinations that are otherwise homozygous *cc* are indistinguishable; they are all *albino* mice. Genes corresponding to those listed here are found in many mammalian species; the synthesis of pigment and control of its distribution within individual hairs and over the body surface are ancient traits. The various terms are described in the text.

Genes						Gametic formula	Phenotype
				P	S	CABDPS	Wild-type agouti
			D		s	CABDPs	Spotted agouti
				p	S	CABDpS	Pink-eyed agouti
		B			s	CABDps	Pink-eyed, spotted agouti
				P	S	CABdPS	Dilute agouti
			d		s	CABdPs	Spotted, dilute agouti
				p	S	CABdpS	Pink-eyed, dilute agouti
	A				s	CABdps	Pink-eyed, spotted, dilute agouti
				P	S	CAbDPS	Cinnamon
			D		s	CAbDPs	Spotted cinnamon
				p	S	CAbDpS	Pink-eyed cinnamon
		b			s	CAbDps	Pink-eyed, spotted cinnamon
				P	S	CAbdPS	Dilute cinnamon
			d		s	CAbdPs	Spotted, dilute cinnamon
				p	S	CAbdpS	Pink-eyed, dilute cinnamon
					s	CAbdps	Pink-eyed, spotted, dilute cinnamon
C							
				P	S	CaBDPS	Black
			D		s	CaBDPs	Spotted black
				p	S	CaBDpS	Pink-eyed black
		B			s	CaBDps	Pink-eyed, spotted black
				P	S	CaBdPS	Dilute black
			d		s	CaBdPs	Spotted, dilute black
				p	S	CaBdpS	Pink-eyed, dilute cinnamon
	a				s	CaBdps	Pink-eyed, spotted, dilute black
				P	S	CabDPS	Brown
			D		s	CabDPs	Spotted brown
				p	S	CabDpS	Pink-eyed brown
		b			s	CabDps	Pink-eyed, spotted brown
				P	S	CabdPS	Dilute brown
			d		s	CabdPs	Spotted, dilute brown
				p	S	CabdpS	Pink-eyed, dilute brown
					s	Cabdps	Pink-eyed, spotted, dilute brown
c plus any other gene						c . . .	Albino, may be of any of 32 genotypes above

color, in *albinism*. The effect of *c* extends to the eye as well as to the hair; in the absence of eye pigment, the red of hemoglobin becomes visible.

The A *locus.* The dominant allele, *A*, is responsible for the *agouti* pattern of pigmentation on individual hairs; this is the coat color that is characteristic of wild rabbits, mice, and rats. Despite being primarily black, each hair has a yellow band more or less midway between its tip and its base. Animals that are homozygous for the recessive allele (*aa*) have solid-colored hairs. Note that the formation of a band on what otherwise would be a solid-colored hair poses an intriguing problem for developmental geneticists. An excellent popular account of the biochemical basis for the agouti pattern was given in the *New York Times,* May 4, 1993. The peptide that is synthesized by the agouti gene (*A*) blocks the receptor of the melanocyte-stimulating hormone. Normally, that hormone stimulates the melanocyte to produce a black pigment (in *BB* or *Bb* animals; otherwise brown; see below); when the hormone is excluded by the agouti polypeptide, only a yellowish pigment is incorporated into the developing hair. The on-and-off timing of the agouti gene remains to be explained.

The B *locus.* This locus is responsible for the presence of black pigment; *bb* individuals are brown. The *A* allele restricts the pigment to the bases and tips of hairs: agouti if black; cinnamon if brown. Animals homozygous *aa* are either black or brown. That there are black and brown bears is a matter related to this account of pigmentation.

The D *locus.* The alleles at this locus determine the intensity of pigmentation. Normal intensity (*DD, Dd*) is dominant to *dilute* (*dd*).

The P *locus.* The effect of the alleles at this locus is expressed in the animal's eye: *PP* and *Pp* individuals have dark eyes; *pp* animals have pink eyes (revealing the eye's blood supply).

The S *locus.* Alleles at this locus determine whether pigmentation is distributed more or less uniformly over the entire body (*SS* and *Ss*) or whether the animal (*ss*) is spotted (*piebald*).

Thus, as shown in Table 5-1, the six gene loci discussed above result in 64 genotypes of which 32 are individually identifiable. The list contains only 33 identifying labels, however, because *cc* animals are albinos whose phenotype fails to reveal those of the 32 genotypes that are generated by the alleles at the remaining five loci.

Here, then, is a study of gene action consisting primarily of an accounting of the many ways in which the products of alleles of various loci interact with one another in producing a phenotype. That each of the 32 nonalbino genotypes can be identified indicates in another sense, however, a lack of interaction. One should be aware in any event that many more genes are involved in determining mammalian coat colors than the six that have been listed here. First, for example, there must be hairs that express the various pigments. Furthermore, modifier genes at many other loci affect each of the six that have been listed. This effect can be shown most readily in the case of spotting. Through artificial selection, the extent of spotting on *ss* mice and rats can be extended or reduced. At one time this fact gave rise to a dispute as to whether the *S* locus was changeable and was continually generating new alleles or, alternatively, whether alleles at other loci *modified* the expression of the *ss* (spotted) phenotype. The matter was resolved by noting the greatly increased variation in degrees of spotting among F_2 animals as compared with F_1 or parental strains (see Table 5-2). An increase in the range of variation within the F_2 generation is expected if contrasting alleles at many loci are segregating independently; an "explanation" based on changeability at a single locus (the *s* locus in the mouse) would require the gratuitous assumption that the degree of changeability depends upon the generation being examined.

Sewall Wright, in addition to his fundamental contributions to population genetics, studied for many years the genetics of coat color in guinea pigs. Just as E. B. Lewis made careful observations of the thoracic and abdominal segmentation in the fruit fly, Wright made careful observations of the pigments in the coats of wild-type and mutant guinea pigs. His conclusions (Figure 5-6) were summarized by Beadle (1945).

First, as noted by Garrod (see Chapter 2), the melanins (complex chemical compounds that constitute the pigments in mammals) arise from intermediate compounds stemming from phenylalanine and tyrosine. Second, melanins of three colors can be recognized: yellow, brown, and sepia. Rather than simply providing a descriptive account of the phenotypes of mice with different genotypes, Wright attempted to explain the effects of these genes in terms of the enzymes produced by their wild-type alleles. The alleles *S* and *s*, which are omitted from Figure 5-6, determine what portions of an animal's skin will exhibit pigmentation; these spotting genes are more concerned with

Table 5-2. The quantitative genetics of coat-color spotting in mice. The parental lines had been subjected to artificial selection for few white patches (lines 118 and 190a) or, alternatively, for large areas of white (line 19) for many (7–17) generations. The F₁ interline hybrids have white areas that, on average, are intermediate to those of their parents in size. The F₂ generations have white spots that, on average, compare with those of the F₁ hybrids. Notice, however, that the distributions of the amount of white on individual F₂ mice greatly exceed those of the corresponding F₁ mice. This increase in *variance* reveals that the degree of spotting is controlled by independently segregating modifier genes not by an ever-changing *S* gene. (After Dunn and Charles, 1937.)

	Percent of dorsal white																				
	00	05	10	15	20	25	30	35	40	45	50	55	60	65	70	75	80	85	90	95	100
P₁ line 118 (F₇₋₁₄)	⋮	07	89	35	2	1															
P₁ line 19 (F₇₋₈)	⋮																	1	9	29	134
F₁ line 19 × line 118	⋮				1	1	4	3	3	10	7	9	9	7	1	1					
F₂ line 19 × line 118	⋮		1	2	6	15	10	12	14	19	27	29	28	19	34	21	14	2	3	2	
P₁ line 190a (F₁₀₋₁₇)	⋮		3	10	22	23	31	27	28	15	5										
P₁ line 19 (F₇₋₈)	⋮																	1	9	29	134
F₁ line 190a × line 19	⋮										1	1	1	5	4	1					
F₂ line 190a × line 19	⋮								2	5		7	16	29	45	10	3	2	3	2	

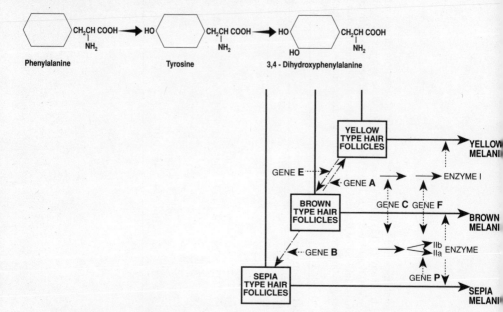

Figure 5-6. A scheme of pigment synthesis based on painstaking observations of the coat colors of guinea pigs of various genotypes (see Table 5-1 for a listing of genes involved in determining the coat colors of mammals). This is an excellent example of the inferences regarding gene action that were necessary before the development of biochemical genetics. (Adapted with permission from Beadle, 1945. Copyright 1945 American Chemical Society.)

cellular differentiation than with the biochemistry of pigment formation.

In the diagram, gene *A* is shown as converting brown-type hair follicles into yellow-type ones; thus, it is a switch gene that is responsible for the occurrence of a yellow band on an otherwise brown- or black-pigmented hair: the agouti pattern. The switch, as we saw above, is the agouti polypeptide that blocks the melanocyte-stimulating hormone.

Gene *C* is positioned in the biochemical pathway in such a manner that, if it fails to function, all pigment formation ceases; the result is an albino animal.

The importance of Figure 5-6 is not in its details, many of which may prove to be wrong. Its importance lies in Wright's attempt at understanding hidden processes of gene action through inferences based on the end results of those processes—both normal and mutant results. The word *inference* is used here in its most respected sense, but with the understanding that in many instances end results are insufficient to identify which of many possible underlying processes are (or is) responsible for them. Biochemical pathways are best revealed by a combination of chemical and genetic analyses, not by an examination of external morphologies; the techniques needed for molecular analyses have only recently been developed.

Drosophila eye pigments

Mutant alleles frequently lead to abnormal pigmentation. Uncritical "explanatory" expressions are frequently used in describing such events; for example, "the *vermillion* mutation in *Drosophila melanogaster* causes bright red eyes." The verb *cause* is an improper one in this case; by blocking a biochemical pathway, the *vermillion* (v) allele *results* in an abnormal eye color, one that is bright, rather than dark, red. These points were addressed by Muller (1932), who added five useful terms to the geneticist's vocabulary; *hypomorph, amorph, hypermorph, antimorph,* and *neomorph.* The first two of these are especially useful.

Eosin (w^e) is one of the mutant alleles at the *white* locus of *D. melanogaster; white,* it may be recalled, was the first (sex-linked) mutant discovered by the Morgan school of *Drosophila* geneticists. Males hemizygous and females homozygous for *eosin* have slightly pigmented eyes. Males exhibiting the *eosin* phenotype have paler eyes than do females homozygous ($w^e w^e$) for the mutant allele. This fact is interesting in itself: most sex-linked eye-color mutants produce the same phenotype in hemizygous (1 dose) males and diploid (2 doses) females. The identity of the two sexes despite the differences in numbers of alleles had earlier led Muller to propose "dosage compensation," that is, to postulate that the single X chromosome in males is physiologically activated so that the single copies of sex-linked genes in males are about twice as active as are the corresponding genes on each of the two X chromosomes in females.

When Muller altered the number of w^e alleles in male and female flies, he found that the additional copies made the flies' eyes more nearly normal in color. Males carrying two copies of the *eosin* mutation resembled diploid females in the color of their eyes; an extra (third) copy of the gene in females made their eyes still darker. Similar observations were made using *apricot* (w^a), *scute*-1 (sc^1), and *forked* (*f*); *scute* and *forked* affect the number and shape of bristles, respectively.

Because extra copies of mutant genes make their carriers more normal (*not* more mutant appearing), Muller suggested that these mutants were not doing something different than normal; rather, they were carrying out the same processes as normal alleles but less efficiently. Increasing the number of these less efficient (*hypomorphic*) alleles caused the phenotype to approach that which is regarded as normal.

Extra copies of many mutant alleles do not tend to restore the normal, or wild-type, phenotype. These are assumed to have lost all function; they are *amorphic*. Among amorphic mutants are *white* (*w*) in *D. melanogaster* and the recessive allele (*c*) that causes albinism in mammals.

Having argued that many mutant alleles are hypomorphic, and recognizing that many mutant alleles undergo "back" mutations (reversions) that make the carriers of the new mutation more nearly normal in appearance, Muller was forced to postulate the existence of *hypermorphic* alleles. Each back mutation is hypermorphic with respect to its hypomorphic progenitor. Others, then, must be hypermorphic to the normal allele. He even found an example: Mutations at the *Notch* locus are generally recessive lethals that, when heterozygous, cause the partial loss of the marginal wing vein. *Abrupt* (*ab*) is a locus whose recessive mutant alleles tend to shorten the interior veins of the fly's wings. However, *abrupt* flies tend not to exhibit *Notch* wings; for this reason Muller suggested that *abrupt* is a hypermorph.

Antimorphic alleles, according to Muller, are those that contribute something other than do their normal (wild-type) alleles when added to an individual's genome. Consider, Muller argued, a fly that, by virtue of experimental manipulation, has the genotype *e/e/+* (*ebony* [*e*] is an autosomal recessive allele that leads to dark body color). The removal of one *e* (thus leaving the heterozygote, *e/+*, results in a lighter body color. Removal of the wild-type (+) allele (thus leaving *e/e*) leads to a darker color. Hence, *ebony*-mutant alleles appear to contribute something that differs from the contribution of the wild-type one.

Neomorphic alleles were postulated as those for which the normal allele behaves as an amorph. *Hairy wing* (*Hw*) mutants add small bristles along the wing veins and on the head and thorax of the fly. Homozygous (*Hw/Hw*) females have twice the number of extra hairs as do heterozygous (*Hw/+*) females or as *Hw* males (*Hw* is a sex-linked mutant). If a wild-type (*Hw⁺*) allele is added to *Hw/+* or *Hw/Hw* females there is no lessening in the degree of hairiness. If a *Hw* allele is added to a *Hw⁺/Hw⁺* fly, extra hairs are formed. The wild-type allele of *Hairy wing* appears to be contributing nothing to the hairy-wing phenotype. At this point it is important to recall that during the 1930s, and even much later, the concepts of genetic activation and suppression were poorly developed, even by someone as creative as H. J. Muller.

Larval eye tissue transplantation in *Drosophila*

Beadle and Ephrussi (1936) carried out a most imaginative study of gene action with respect to eye color in *Drosophila—the* most imaginative of that era. Indeed, Beadle was to write later (1963, p. 13) that the phrase "one gene–one enzyme" was coined in conjunction with these experiments on *Drosophila,* even earlier than its popularization following the biochemical studies using *Neurospora.*

An understanding of *Drosophila* larval development is needed in order to appreciate Beadle and Ephrussi's experiments. Upon hatching, the *Drosophila* egg yields a tiny, first-instar larva. Eating vigorously in a well-yeasted culture at 25°C, the first-instar larva sheds its skin and chitinous mouth parts within 25 hours, at which time it becomes a second-instar larva. A second molt occurs at the 48th hour, thus yielding the third-instar larva. Two days later, from the 96th until the 100th hour, the third-instar larva becomes sluggish, forms a puparium, undergoes a prepupal molt, and completes pupation by the 108th hour. At age 192 hours (8 days), the adult fly (*imago*) is ready to emerge from its pupal case.

Internally, future adult tissues and organs are laid down in the growing larva. An eye-antenna complex of cells is visible in the first-instar larva within an hour after hatching. By the 50th hour, early in the third-instar larva, the eye-antenna disc has become clearly differentiated into two portions, eye and antenna; both portions grow steadily in size through the next two days that lead to pupal formation.

Ephrussi and Beadle utilized larva of different genotypes either as donors of eye discs or as hosts for transplanted eye discs. Their technique required not only fine microinjection pipettes and a microinjection apparatus but also a great deal of manual skill. Each host larva, for example, had to survive, continue its development, pupate, and emerge as an adult. At that time, the fate of the transplanted eye disc could be determined because it now resided as an adult eye in the host fly's abdomen. The outcomes of several of Beadle and Ephrussi's experiments are illustrated in Figure 5-7. At the left are listed the genotypes of larvae from which transplanted eye discs (implants) were obtained. Across the top are listed the genotypes of the host larvae. Shaded circles indicate that the implanted eye discs developed the same eye color as the mutant strains from which they were obtained. For example, the shaded circles lying on the diagonal leading from the upper left to the lower right reveal that the experimental manipulations themselves had no effect on eye color. Elsewhere in the diagram, the shaded circles indicate that the transplanted eye discs were autonomous in that the adult eye color of the transplant was that corresponding to its genotype rather than otherwise (for example, that of its host). Eye discs obtained from *scarlet* (*st*) larvae and transplanted into *sepia* (*se*) larvae, to cite one example, became bright red (i.e., scarlet) eyes lying within the abdomens of sepia-eyed adult flies. The solid circles in Figure 5-7 reveal more interesting events than do the shaded ones. They indicate transplantations in which the implanted eye disc took on a wild-type eye color (*not* the eye color of a mutant host!). An exception occurs in the claret-like color of wild-type discs that were implanted into claret flies. To anticipate, we can suggest here that the *claret* host had insufficient material of some sort to "support" the normal development of eye color in the wild-type implant. Thus, *cinnabar* implants in wild-type, *cardinal*, and *sepia* hosts developed into wild-type eyes. *Vermillion* implants yielded wild-type eyes in wild-type, *cardinal, cinnabar,* and *sepia* hosts. The most interesting of the results illustrated in Figure 5-7 involve implants derived from *cinnabar* and *vermillion* donor larvae. Transplanted into wild-type larvae, both types of mutant eye discs yielded wild-type adult eyes. The wild-type host larvae provide something that enables these mutant eyes to develop normal color. (Recall that when transplanted into hosts of the same mutant genotype—*v* into *v* and *cn* into *cn*—the mutant eye discs retained their

HOST

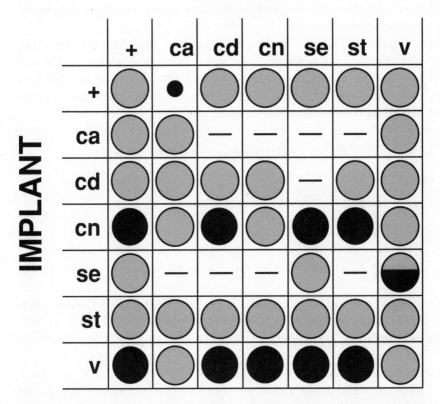

Figure 5-7. Diagram summarizing much of the data from Beadle and Ephrussi's (1936) experiments on the transplantation of larval eye discs between wild-type and mutant individuals. The mutant (or wild-type) donor strain is listed at the left; the recipient (host) type is listed at the top. Shaded circles indicate that the transplanted disc gave rise to an adult eye whose color was that of the original strain (autonomous coloring). Solid circles indicate that the transplanted disc gave rise to an adult eye whose color differed from that of the donor strain. The large solid circles indicate that the transplanted eye became wild-type in color; the single small solid circle indicates that, when implanted into *claret* flies, the wild-type implant was claret in color. *Sepia* implants within *vermillion* larvae (the half-shaded, half-solid circle) gave intermediate results that were also sex dependent. *ca, claret; cd, cardinal; cn, cinnabar; se, sepia; st, scarlet; v, vermillion.* Dashes represent tests that were not performed.

mutant, bright red color.) On the other hand, whereas *vermillion* eye discs became wild-type in color when transplanted into *cinnabar* hosts, *cinnabar* eye discs remained *cinnabar* when they developed within *vermillion* hosts.

Beadle and Ephrussi provided the correct interpretation for these observations; the normal alleles at the *vermillion* and *cinnabar* loci provide the means (enzymes) by which one diffusible substance is converted into a second one, which is then converted into a third according to the following scheme:

$$A \xrightarrow[v^+]{} B \xrightarrow[cn^+]{} C \longrightarrow \longrightarrow \text{Pigment}$$

Subsequently, the hypothetical substances (A, B, and C) were isolated and identified:

tryptophan $\rightarrow$ formylkynurenin $\rightarrow$ hydroxykynurenin $\rightarrow$ $\rightarrow$ pigment

Thus, *vermillion* mutants, which are unable to synthesize formylkynurenin, are able to convert that substance (which diffuses into the implanted *vermillion* eye from the *cinnabar* host) into hydroxykynurenin and, eventually, pigment. The *vermillion* eye that is implanted into a *cinnabar* larva develops a wild-type color. Conversely, a *cinnabar* eye implanted into a *vermillion* larva remains *cinnabar* in appearance. The *cinnabar* eye is unable to transform formylkynurenin into hydroxykynurenin, and the *vermillion* larva is unable to synthesize even formylkinurenin.

These studies by Ephrussi and Beadle were among the first to skillfully pose the question, What do genes do? Eventually, the question was approached productively by using different experimental organisms. *Neurospora* (a bread mold) and *Escherichia coli* (a colon bacillus) normally synthesize virtually all of their chemical components from water, sugar (an energy source), some inorganic salts, and one (or a few) vitamins. These organisms offer the geneticist an opportunity to interfere with nearly every one of the biochemical steps needed for maintaining life. The first experiment in this exciting direction, however, was taken by noting the pigmentation of transplanted *Drosophila* eyes.

Plant colors

Plant pigments, especially those of flowers, provided early geneticists with splendid markers for studying patterns of inheritance. Not only did these studies reveal the frequent incompleteness of dominance (a consistent phenomenon in Mendel's studies) but they also revealed numerous (epistatic) interactions between genes at different loci. Small wonder, then, that these plant pigments were regarded as offering keys to an understanding of gene action. Haldane (1942), in his book *New Paths in Genetics* devoted considerable space (pp. 54–82) to plant pigments and their interrelations. Beadle (1945, pp. 35–40) also reviewed the "anthocyanins and related plant pigments." Haldane concluded his chapter with this observation (p. 82): "We see that a geneticist cannot possibly neglect biochemistry. I hope that I have shown that the study of genetics is not without value to the biochemist." The observation is certainly proper. However, in both reviews, the known facts are primarily those concerning the organic structures of plant pigments; no sequence of enzymatic synthesis comparable to the Ephrussi-Beadle observation emerged from the study of plant pigments. Rather than concentrating on metabolic pathways, the early studies on plant pigments appear to have dealt with differences between the pigments themselves: that is, with static rather than dynamic information. The ultimate in exposing the dynamics of life was to be found in those microorganisms that literally start from scratch in maintaining and reproducing themselves.

A different sort of concern regarding plant pigmentation was that which dealt with the *distribution* of pigmentation on the plant. Two loci in *Zea mays* were especially important in this regard: the R (aleurone and plant color) and A (aleurone color) loci. (The aleurone is the outer cell layer of the corn kernel's endosperm; it is of embryonic origin.) The first, consisting of the four alleles R^r, R^g, r^r, and r^g, was especially important in leading to an understanding of both gene structure and gene function.

The R locus in maize controls anthocyanin pigmentation; it is located at map position 111 on chromosome 3. Provided the proper genetic background (a standard admonition in genetic studies because genes are only *necessary*, never *sufficient*, for the generation of their characteristic phenotypes), the R gene controls the purple pigmentation of both the plant (including anthers and leaf tips) and

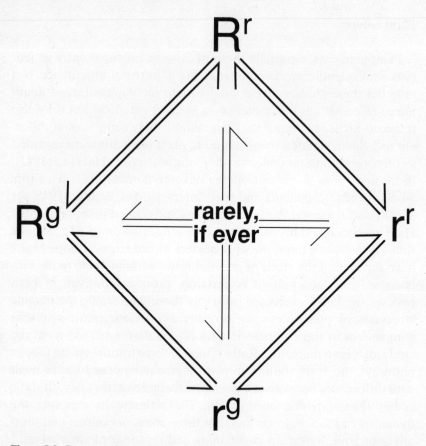

Figure 5-8. Patterns of mutation that alter the *R*-locus in *Zea mays*. This locus governs the distribution of anthocyanin pigment on the plant (*R*, purple plant; *r*, green plant) and on the seed (*r* superscript, purple aleurone layer; *g* superscript, colorless aleurone). Mutations were observed that changed one item (e.g., $R \rightarrow r$ or, among superscripts, $r \rightarrow g$) at a time, but seldom if ever were two changed simultaneously ($R^r \rightarrow r^g$ or $R^g \rightarrow r^r$, for example). Such observations suggested that this locus was occupied by two genes; subsequent studies revealed that it is, indeed, a small duplication of genetic material.

the aleurone layer of the seed (kernel). The pattern of pigmentation depends, however, upon the particular allele: one (R^r) leads to both plant and seed color, a second (R^g) to plant color alone, a third (r^r) to green plants with colored seeds, and a fourth (r^g) to colorless seeds on a green plant (see Figure 5-8).

The four alleles undergo mutations at low frequencies. The allele

R^r mutates to R^g and to r^r. The allele R^g mutates to R^r and to r^g. Similarly, r^r was observed to mutate to R^r and to r^g. Never, apparently, does R^r mutate to r^g or R^g to r^r, or vice versa. Whereas crossing-over had failed to suggest that the R locus was any more than a single gene, mutation studies strongly suggested that two closely linked genes were present at the R locus—one concerned with plant pigmentation, the other with seed color.

Subsequent studies revealed that, in fact, the R locus consists of a small chromosomal duplication. As (bad) luck would have it, the left-hand member of the duplication controls seed pigmentation, and the right-hand member the pigmentation of the plant; this, it can be noted, is the reverse of the earlier notation. Consequently, the symbolism might be changed: SP (S = seed color; P = plant color) = R^r; $Sp = r^r$; $sP = R^g$; and $sp = r^g$.

The realization that the R locus represented duplicated genes (see Emmerling, 1958) and that the A locus was also a duplicated locus (Laughnan, 1955) led to extensive studies dealing with the nature of gene mutations: Are they intragenic, or extragenic? An unassailable consensus of the era held that intragenic mutations were merely a residue that remained after extragenic mutations (small deletions, inversions, and duplications) had been identified as such. The concept of a unitary gene—a bead—still governed the thinking of most geneticists; that the gene, itself, could undergo internal deletions, inversions, and duplications because its physical structure did not differ from extragenic material (that the gene is part of the string, not a bead *on* the string) lay beyond comprehension. The resolution of this misunderstanding awaited the identification of DNA as *the* genetic material and an understanding of the relationship (including colinearity) between the bases of DNA (in triplet sets known as codons) and the amino acid sequences of proteins.

The concluding paragraphs dealing with plant pigmentation and the understanding of gene action outline Barbara McClintock's early studies on unstable gene expression in corn endosperm. Mutable gene loci had long been considered to hold a key to an understanding of gene action. Variegated pericarp (the outer layer of the corn kernel) in *Zea mays* was one of the first unstable phenotypes observed. The larkspur, *Delphinium*, has many varieties with variegated flower colors (Figure 5-9). Unless the variegated portion includes at least a fraction of the plant's gametes, one is restricted to

Figure 5-9. Flowers of *Delphinium,* showing the somatic mutations that alter segments of the blossom from rose color to purple. An early-occurring mutation gives rise to a large altered segment; a late-occurring one to a small purple patch. There appear to be "preferred" times for mutation, early and late; otherwise, one would expect a continuous distribution of sector sizes. (After Demerec, 1931.)

describing the size of the abnormal spots (early mutational events lead to large abnormally pigmented sectors; late mutations to small ones) and to determining their frequencies. Microbial geneticists would remind us that similar sectoring occurs in bacterial and fungal colonies but, in those cases, cells within these sectors can be removed, recultured, and grown in quantity for further genetic or biochemical analysis.

During the early 1940s, Barbara McClintock carried out studies on the breakage and subsequent fusion of "broken" chromosomes in the nuclei of corn plant cells. This study included an analysis of the induction of new mutations: aleurone color, seedling color, plant color, and kernel and leaf shapes. In her report to the trustees of the Carnegie Institution of Washington (1946), she described "the unexpected appearance of a number of unstable mutants": variegated white seedling, variegated light-green, a variegated luteus (yellowish) mutant, and chromosomal-breakage variegation. These studies can be included in a chapter devoted to "color" simply because it was the unstable color patterns (including the green of chlorophyll) that caught McClintock's attention. As she reported in 1946, there appeared (in the case of the luteus locus) to be stable recessive alleles that rarely mutated to dominant ones, stable dominant alleles that rarely mutated to recessive ones, and a variety of recessive and dominant alleles that exhibited intermediate rates of mutability. Similar assertions could, of course, have been made following an examination of somatic color changes in *Delphinium* (Figure 5-9).

The chromosome-breakage variegation first described by McClintock in 1946 will be the focus here. The phenomenon was observed in the progeny of a self-pollinated plant that carried the alleles *I* (dominant inhibitor of aleurone color) and *Wx* (endosperm starch stains blue with iodine) in one chromosome 9, and the alternative alleles, *i* (colored aleurone) and *wx* (starch stains red with iodine) in the other. (In 1946, *I* and *i* were considered to represent a locus that was tightly linked to the *C* locus, the dominant allele of which is responsible for colored aleurone; McClintock's selfed corn plant, consequently, was homozygous *CC*. The modern symbolism would be $C^I Wx$ and $C wx$ for the two chromosomes because the dominant inhibitor is now regarded as one of the *C* series of alleles: C^I [dominant inhibitor of color], *C* [colored aleurone], and *c* [recessive, colorless aleurone].)

Here we will retain the older symbolism. Kernels that received both *I* and *i* should have been colorless because of the dominant (color) inhibitor, *I*. However, some kernels possessed well-defined sectors that exhibited colored spots whose uniformity in both size and distribution within the exceptional sector suggested that the *I*-bearing chromosomal segment was being systematically lost. By coincidence, the *I*-bearing chromosome also carried a small chromosomal deficiency (*wd*) that, when homozygous, causes white seedlings. Many of the colorless kernels from this self-pollinated plant gave rise to white seedlings; i.e., the *Wd* "allele" of the other chromosome was missing.

The subsequent steps in McClintock's analysis of the origin of the observed instability involved sophisticated genetic crosses involving not only *Wd-wd, Wx-wx,* and *I-i,* but also *Bz-bz* (bronze, very light aleurone color) and the cytological examination of pachytene chromosomes of the microsporocytes (early pollen-forming cells) of mature plants. The genetic tests suggested that chromosome 9 (the chromosome on which these genes are located) was systematically losing a terminal portion, with the break occurring close to the *Waxy* locus; the cytological observations confirmed this loss.

In subsequent years, McClintock determined that two genetic elements played major roles in giving rise to her variegated plants and

Table 5-3. Genetic evidence for the transposition of a gene-controlling element (*Spm*) from one location to another within the genome of *Zea mays*. The test involved 44 ears obtained from 17 plants. Forty-three ears gave similar results and are combined in the first line; these data reveal that *y* is not associated with *Spm*, whereas *Y* is associated with that element. The remaining ear (one of several produced by a single plant) shows, in contrast, that the inheritance of *Spm* and *Y* is independent. In this ear, *Spm* has physically moved and is no longer genetically linked to *Y*. (After McClintock, 1957.)

	Pale color (no *Spm*)		Colorless (*Spm*)		
	Y	y	Y	y	Total
43 ears	273	5002	4742	204	10,221
1 ear	65	47	48	59	219

kernels: *Ds* (*Dissociator,* the element that caused the chromosome breaks) and *Ac* (*Activator,* an element whose presence was needed in order to "activate" *Ds*). This portion of McClintock's research will be discussed in Chapter 9, which deals with transposable chromosomal elements. Here we shall dwell upon observations that revealed the transpositions in the chromosomal position of *Ds* and Ac; these observations were the ones that, even though scrupulously supported by data, roused the skepticism of many geneticists. In her annual report, McClintock (1948) described chromosomal breakages that occurred coincidentally with ones caused by *Ds;* fusions may then occur that give rise to gross chromosomal aberrations—translocations, for example. These aberrations can shift the position of *Ds* in the genome; one particularly useful shift moved *Ds* from near *Wx* (position 9-59) to a locus between *I* and *Sh* (9-26-9-29). "Useful" because of the additional observations it permitted McClintock to make regarding chromosomal breakage, observations that confirmed her cytological observations. *Ac* also moved about in the genome. Many of McClintock's early crosses revealed the independent assortment of *Ac* and *Ds*—many, but not all. In three cases, *Ac* proved to be linked to *Ds* and about 6–20 crossover units to the right. Data related to a shift in the physical location of still another "controlling element" (*Spm,* suppressor and mutator) are shown in Table 5-3, where it is seen that in one of several ears produced by a single plant, *Spm* moved from a position that was about 5 crossover units from the *Y-y* locus (6-17) to an entirely different chromosome, thus exhibiting independent assortment among the kernels on one ear of corn.

Recapitulation

Warn your audience what you intend to say, say it, and then review what you have said. Those were the advisory statements made earlier in this chapter. The intent has been to reveal the role that color (or organic pigments) has played in the study of gene action. The most important role in our opinion has been that of an indicator; thus, aside from their visibility, it was mere accident that purple, bronze, and colorless aleurones as well as white and green seedlings enabled McClintock to untangle the inheritance of and variation in gene control elements—elements that alter the expression (including complete inactivation) of both normal and mutant alleles. It was also accidental that levels of pigmentation allowed Muller to infer that many mutant alleles do precisely what their wild-type alleles do, only less efficiently. On the other hand, the chemistry of pigments has interested those who set out to untangle the intricacies of metabolic pathways; in this case, the enzymes that are responsible for the necessary chemical transformations have often been of less interest than the complex chemicals on which they operate. In brief, however, one must confess that genes—no matter what aspect of the phenotype on which they impinge—act primarily by way of the enzymes that they specify. This is equally true for differences in morphology and in color.

6   Position Effect

The first instance in which it was observed that the position of a gene within an organism's genome affected its functioning was the *Bar* mutation in *Drosophila melanogaster*. A chapter on *position effect* (i.e., the effect of its position within the chromosome on a gene's action) might well begin with a review of this mutant and the studies of which it was an integral part.

The mutant was first seen in 1914 as a small-eyed male (Figure 6-1B). Standard genetic tests quickly revealed that the gene was sex linked and was located at map position 57 of the X chromosome. Figure 6-1 depicts the phenotypes of (A) a wild-type fly, (B) a *Bar* male or homozygous *Bar* female, and (C) a female heterozygous for the *Bar* mutation ($+/B$). The small, rather triangular indentation (somewhat obscured by the fullness of the eye itself, thus producing a kidney-shaped eye), is characteristic of *Bar* heterozygotes, which, because the trait is sex linked, must be female. *Bar,* consequently, exhibits partial dominance.

Within five years of its discovery, *Drosophila* workers had noted that homozygous *Bar* (B/B) strains of flies are unstable; about one son of 1600 produced by a B/B female is wild type in appearance, and a similar number have eyes that are even smaller than *Bar*-eyed flies (Figure 6-1D). These observations were converted into an "explanation": *Bar* is an unstable mutant that frequently reverts to wild type or mutates to a more extreme allele of *Bar* (*Double-Bar* or *Ultra-Bar*). Note that this statement is *not* an explanation; it merely describes the observations.

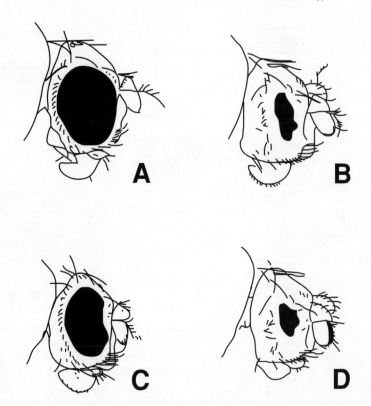

Figure 6-1. The relative sizes and shapes of the compound eyes of (A) wild-type, (B) homogyzous or hemizygous (male) *Bar,* (C) heterozygous *Bar* female, and (D) homozygous or hemizygous *Ultra-Bar* (also known as *Double-Bar*) *Drosophila melanogaster.*

In 1936, nearly 20 years after the early studies of the *Bar* mutation, both H. J. Muller and C. B. Bridges discovered by an examination of the giant salivary gland cell chromosomes of *D. melanogaster* that the *Bar* mutation is in reality a small chromosomal duplication (Figure 6-2). The bands that fall within the cytological region identified as 16A1–7 in the wild-type or normal X chromosome are duplicated in the chromosome, causing the *Bar*-eye condition. Furthermore, the chromosome that is responsible for the *Double-Bar* phenotype contains three copies of the region 16A1–7. The instability of the homozygous (*B/B*) *Bar*-eyed strain of flies now becomes clear: crossing-over within the duplicated region when the two chromoso-

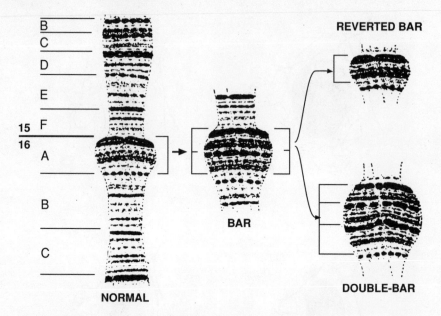

Figure 6-2. The *Bar* region of the X chromosome of *Drosophila melanogaster*. In *Bar*-eyed flies, section 16A1–7 is duplicated. By "unequal" crossing-over in homozygous *Bar* females, chromosomes can either lose the duplicated section and revert to wild type (*Reverted Bar*) or pick up an additional copy of segment 16A1–7 and become *Double-* (or *Ultra-*) *Bar*.

mal segments have mispaired, as shown in Figure 6-3, leads to the production of two daughter chromosomes—one with a single segment (= wildtype) and the other with three (= *Double-Bar*).

With this introduction to the formal genetics of the *Bar* phenotype, we can now turn to the matter of *position effect,* or the effect of position on genes. The instability of homozygous *Bar* strains has already suggested that no irreversible change occurred during the change from wild type to *Bar,* otherwise no revertants to wild type could be obtained. They are obtained, however, thus proving the point. The three phenotypes—wild type, *Bar,* and *Double-Bar*—could result from having 1, 2, and 3 "doses" of the chromosomal segment, 16A1–7.

Sturtevant pointed out that the situation is more complex: Wild-type flies have (on average) about 750 facets in their compound eyes; homozygous *B/B* females about 60; heterozygous, *B/+*, females about 360; and homozygous *Double-Bar* (*BB/BB*) females about 25. (These numbers can be compared with the drawings in Figure 6-1.)

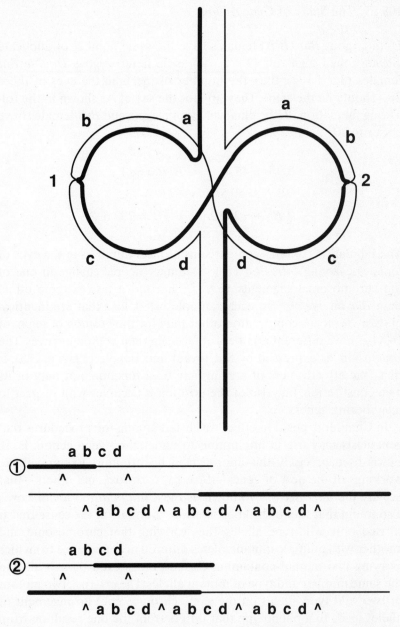

Figure 6-3. A diagrammatic representation of the "unequal" crossing-over that leads to *Bar* revertants and to *Double-* (or *Ultra-*) *Bar*. The carets demark the section *abcd* that corresponds to section 16A1–7 of the salivary chromosome. Two areas within which crossing-over may occur are designated 1 and 2; their crossover products are shown in the lower diagrams.

Homozygous *Bar* (*B*/*B*) females have the same number of duplicate chromosomal segments (2 + 2 = 4) as do heterozygous *Double-Bar* females (3 + 1 = 4); thus, the number of facets in the eyes of these flies should be the same. They are not the same! As shown in the following tabulation, *BB*/+ females have 45 facets and *B*/*B* females have closer to 70.

$$BB/+ = 45.4 \qquad B/B = 68.1$$
$$BB^i/+ = 50.5 \qquad B/B^i = 73.5$$
$$B^iB^i/+ = 200.2 \qquad B^i/B^i = 292.6$$

The tabulation contains comparisons of facet numbers in the eyes of females carrying *Infra-Bar* (*B*^{i}), a submicroscopic change in one of the chromosomal segments 16A1–7 that has a less extreme effect than *Bar* on eye size. In each example listed, flies that are identical in their *Bar* locus composition other than for the *position* of segment 16A1–7 have different numbers of facets in their compound eyes. The conclusion, as expressed by Sturtevant and Beadle (1939, p. 225), is that "the effectiveness of a gene may be a function, not only of its own constitution, but also of the position it occupies with respect to neighboring genes."

In Chapter 4, position effect was listed among the procedures that geneticists may use in attempting to understand gene action. E. B. Lewis used precisely this approach in his attempts to unravel the workings of the nest of genes—*Ubx, bx, Cbx, bxd,* and others—that occupy the *bithorax* (now known as the *Ultrabithorax*) locus. Lewis also found that if one chromosome was "normal" in the sense that it carried only wild-type alleles, flies carrying that chromosome and another with multiple mutant alleles differed in appearance from flies carrying two mutant-containing chromosomes even though all had the same number and type of mutant alleles. The *cis-* (i.e., two mutant or two wild-type alleles on the same chromosome) arrangement of alleles leads to a phenotype that differs from the one resulting from the *trans-* arrangement of the same alleles (i.e., one mutant and one wild-type allele on one chromosome, and the other mutant and wild-type allele on the homologous chromosome). *Cis-* and *trans-* arrangements eventually became extremely important in the study of microbial and phage genetics both for defining the gene ("cistron") and in studying gene regulation. In higher organisms, using mutant genes removed from one another by several map units or more, *cis*- and *trans*- arrangements (or, using earlier but corresponding terms: *coupling* and *repulsion*) have little, if any, effect on the ultimate phenotype.

The mechanism of position effect: An early consideration

Boris Ephrussi and Eileen Sutton (1944), two geneticists at Johns Hopkins University, prepared one of the first reviews in which information concerning position effect was subjected to interpretation. That review presented a clear statement of the two prevailing views regarding the causes of position effect at the time and, therefore, is worth citing. In addition, however, it offers clues concerning the politics or sociology of science; these will be touched on as a digression in the following section. In justifying the effort to understand position effect, the authors admit the restriction of well-established cases to a single species, *Drosophila melanogaster;* however, in this species, it proved to be a much more common phenomenon than originally suspected. In brief, one might argue that the greater detail in which the genetics of the fruit fly was known, the more likely one was to encounter a particular phenomenon—and the more tempting it became to concentrate on those phenomena still awaiting explanation. Subsequent to their review, D. G. Catcheside (1939) observed a case of position effect in the plant *Oenothera blandina*.

How might two genes that are brought into a novel proximity interact with one another and, thus, come to express themselves (one, the other, or both) in an unusual fashion? The first explanation claimed that the products of genes diffuse from the gene into the protoplasm, where, by interacting with other substances, they produce their effects. When brought into an unusual proximity, the products of two genes may undergo unusual reactions with one another. This interaction of gene products may interfere with the subsequent, proper, interactions of either with other cellular components. This became known as the *kinetic* hypothesis. It has an inverse formulation: when two genes have been brought into proximity, their products compete for a common substrate. The outcomes of these two formulations in terms of the kinetics of chemical reactions are the same: the interaction of two products with one another mimics the competition of the two products for the same (and limiting) substrate.

The second hypothesis became known as the *structural* one. This hypothesis postulates that the unusual proximity of two genes modifies the structure of the genes. Just as, for example, the addition of a chromophore to a large organic molecule may convert it into a colored compound (or alter its original color), so might the placement of an "alien" gene next to a "native" one alter the physical properties of either or both. The proponents of this hypothesis were immediately confronted with the observation that not all cases in which genes were brought into unusual proximities gave rise to position effects; one could only admit that "some do, some don't."

Ephrussi and Sutton (1944, p. 188) came to favor the following view: "The factor primarily responsible for position effect corresponds to a change in the physical state of the chromosome itself rather than to a change in the distribution of a substance emanating from it." Their preference was for a modified structural hypothesis, not the kinetic model. To recite the reasons why Ephrussi and Sutton favored a model that, in today's perspective, is necessarily crude is unnecessary. Nevertheless, one observation that was crucial to their decision was the following (p. 188): "Position effects are now known to extend over such great distances that it is difficult to see, on the basis of the kinetic hypothesis, why interactions similar to those which occur between two given genes in position effect should not occur between the same genes in their normal location."

The critical observations for the Ephrussi-Sutton hypothesis were those of Demerec (1940). Before presenting Demerec's observations, we must make a clarifying point: There are two broad categories of position effect, stable (S) and variegated (V). *Bar* eye is an example of S position effect; the phenotype (small eye in the case of males and homozygous females; kidney-shaped eye in the case of heterozygous females) is stable in the sense that it can be relied upon to characterize the progeny flies in mutant strains (except for the origin of wild-type and *Double-Bar* chromosomes through "unequal" crossing-over) generation after generation.

Variegated position effect is so named because, when euchromatic material is brought into an unusual proximity to heterochromatin (especially of centromeric regions), wild-type genes may behave as

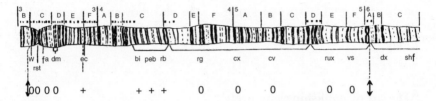

Figure 6-4. The outcome obtained by inserting a 150-band segment of the X chromosome into the highly heterochromatic chromosome 4 (Notch 264–85). The wild-type genes at the *white* (*w*), *roughest* (*rst*), *facet* (*fa*), and *diminutive* (*dm*) loci on the left and at the *rugose* (*rg*), *curlex* (*cx*), *crossveinless* (*cv*), *roughex* (*rux*), and *vesiculated* (*vs*) loci on the right exhibit mottled (variegated) expression (represented by zeroes), whereas the wild-type alleles at the four loci in the center of the insert— *echinus* (*ec*), *bifid* (*bi*), *pebbled* (*peb*), and *ruby* (*rb*)—remain unaffected (represented by plus signs) by position effect. Notice that the variegated and nonvariegated loci are not interspersed; the suppression of gene action "invades" from either end of the insertion (see Figure 6-11). (After Demerec, 1940, by permission of *Genetics*.)

their mutant alleles in certain cells while remaining wild type in others. Thus, a wild-type fly whose *white* locus has been placed near heterochromatin by either a chromosomal inversion or translocation may develop a mottled eye: red in many facets, as it should be, but mottled by the occurrence of many white areas (ranging in size from one to dozens of facets) in which the w^+ allele seems to have been inactivated, thus behaving as *w*, the amorphic allele.

Demerec obtained by means of X rays a number of *insertions*, segments of the euchromatic portion of the X chromosome that were removed by means of two chromosomal breaks and that were then incorporated into the heterochromatic position of an autosome, either chromosome 3 or 4. He saw that, if the phenotype of a given wild-type allele at any locus showed mottling, those of all known loci between that locus and the "alien" heterochromatin also showed mottling. This finding was true for insertion N264–85, where mottling occurred at gene loci occupying both ends of the inserted enchromatic portion (Figure 6-4). Interestingly, Ephrussi and Sutton (1944, p. 188) stated "that the explanation of the mechanism of position effect must be sought in terms of a *factor which will spread along or across the chromosomes rather than through the karyolymph* [emphasis theirs]." In abandoning the kinetic hypothesis, they opted for one based on physical (mechanical?) change; thus, they neglected the factor that might "spread along . . . the chromosome," a possibility that would include the transcription of DNA as it is understood today.

"The art of publishing obscurely"

The purpose of this digression is to reveal an interesting facet of the sociology of science that might otherwise be overlooked in the Ephrussi-Sutton review article. "The Art of Publishing Obscurely" is the title of an essay by Garrett Hardin (1964, p. 116) in which he discusses the characters who emerged seemingly from nowhere to claim prior credit once Darwin had published his *Origin of Species* in 1859. A Patrick Matthew, in the *Gardener's Chronicle* for April 7, 1860, claimed credit for the idea of natural selection; he had published on the subject in 1831—in an appendix to an article on naval timber and arboriculture.

In introducing the kinetic theory, Ephrussi and Sutton stressed the contribution of Muller to the discussions of position effect "in terms of diffusion." (The title of Muller's 1938 paper is "The Position Effect as Evidence of the Localization of the Immediate Products of Gene Activity.") This stress was underscored subsequently when Ephrussi and Sutton introduced the structural theory by pointing out (1944, p. 185) "that the difficulties of the structural hypothesis have been from the very first so fully realized by its author [H. J. Muller] that he definitely gave preference to what he called the kinetic hypothesis." And, indeed, Muller (1947) did say (p. 27), "It is conceivable that these 'position effects' are due to localized chemical reactions between the products of nearby genes. But it seems more likely that they are caused by changes in shape of the gene, that give it a different amount or even direction of effectiveness." This sentence terminates with a reference to an earlier, 1935, paper.

In the words of Garrett Hardin near the end of his essay: "One of the most capable geneticists of the twentieth century made life miserable for his colleagues by his all too frequent use of the Ploy of the Significant Footnote." The ploy being, of course, to come out on both sides of an unresolved issue so that, following its resolution, reference can be made to one statement while the other goes unnoticed. Ephrussi and Sutton were well aware of this ploy; hence, their exag-

gerated references to Muller, emphasizing that he was a key "propo-
nent" of *both* the kinetic and the structural hypotheses of position
effect.

Sutton's analyses of *Bar* revertants

In 1943, Eileen Sutton published the results of a study of the *Bar*
"locus" in which she employed a procedure that, even today, is one
of the more powerful analytic techniques: namely, the induction of a
particular mutation by a mutagenic agent (merely to speed up the
analysis), followed by the induction of the wild-type phenotype as
revertants from the mutant individuals. The process, of course, can be
repeated; the wild-type revertants can be re-mutagenized and these,
in turn, can be induced to revert to wild type once more. Biochemical
or molecular geneticists sometimes refer to this procedure as "three-
dimensional" genetics as opposed to "linear" genetics, in which the
focus is on the linkage relationships of gene loci. In three-dimensional
genetics, one presumably alters the gene product (a protein). A wild-
type revertant of that mutant may be either a true reversion (restoring the gene to its original base composition) or a compensatory
"reversion" (suppressor mutation). The latter may consist of a second
change within the same protein molecule or a change in a second
protein that restores the interaction between it and the first, now
altered, protein. Both restore the protein's normal function. The
latter case (i.e., the restoration of functional *interactions*) allows one
to identify interacting proteins by a combination of genetic and
biochemical procedures. For example, proof that two proteins are
involved in histidine transportation in *Salmonella* was obtained by
examining "wild-type" revertants of strains carrying a mutation at
either one of two gene loci; the revertants in either case proved in
many instances to have acquired a mutation at the other locus.

Sutton (1943) proceeded in just this manner with the *Bar* locus but
with a technical limitation; she was limited to a cytological examination of giant salivary gland cell chromosomes. First, she examined the
chromosomal changes that accompanied mutations from wildtype to
Bar mutants (Figure 6-5). She analyzed a total of six mutations; four
were already available, and two she obtained herself. Each mutation
is characterized by a chromosomal break that lies to one side or the

+ to B

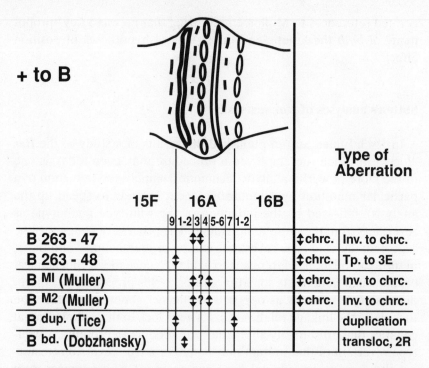

	15F	9	1-2	3	4	5-6	7	1-2	Type of Aberration	
			16A					**16B**		
B 263 - 47			‡‡						‡chrc.	Inv. to chrc.
B 263 - 48	‡								‡chrc.	Tp. to 3E
B MI (Muller)			‡?‡						‡chrc.	Inv. to chrc.
B M2 (Muller)			‡?‡						‡chrc.	Inv. to chrc.
B dup. (Tice)	‡					‡				duplication
B bd. (Dobzhansky)	‡									transloc, 2R

Figure 6-5. The *Bar* region of the X chromosome (Section 16A, bands 1–7) of *Drosophila melanogaster* and the nature of six aberrations that alter the fly's phenotype from normal (wildtype) to *Bar* (B^bd. = *Baroid*). Double-headed arrows indicate points of breakage (chrc. = chromocenter of X chromosome). The associated aberrations are inversion (inv.), transposition (Tp.), translocation (transloc), or duplication. (After Sutton, 1943, by permission of *Genetics*.)

other of the heavy doublet bands that are known as 16A1–2. The second break can be near (as in the original *Bar* of Tice) or far. The common feature accompanying the origin of *Bar* is a change in chromosomal structure (revealed as a chromosomal rearrangement) such that an alien chromosomal band is placed (either proximal or distal) adjacent to 16A1–2.

Every *Bar* mutation revealed the presence of an alien band next to 16A1–2. Does every band that is moved to that position cause *Bar* eyes? Recall that Sutton's screening procedure was to search for *Bar*-eyed mutants, not to search for chromosomal aberrations. The search for wild-type revertants of *Bar* provided the answer (Figure 6-6). One revertant, 263-46, neatly removed one of the duplicated segments as

B to + ➡

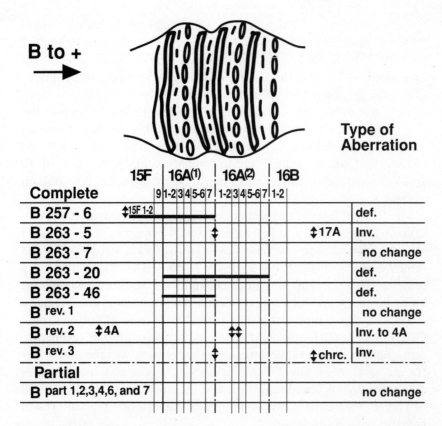

Type of Aberration

Complete	15F	16A(1)	16A(2)	16B	Type of Aberration
	9 1-2 3 4 5-6 7	1-2 3 4 5-6 7	1-2		
B 257 - 6	↕15F 1-2 ▬▬▬▬▬▬▬▬▬				def.
B 263 - 5		↕	↕17A		Inv.
B 263 - 7					no change
B 263 - 20	▬▬▬▬▬▬▬▬▬▬▬				def.
B 263 - 46	▬▬▬				def.
B rev. 1					no change
B rev. 2 ↕4A		↕↕			Inv. to 4A
B rev. 3		↕	↕chrc.		Inv.
Partial					
B part 1,2,3,4,6, and 7					no change

Figure 6-6. The *Bar* duplication of the X chromosome of *Drosophila melanogaster* and the nature of chromosomal alterations that restore the wild-type phenotype. In addition to the aberrations listed for Figure 6-5, restoring the wild-type phenotype in several instances involves the loss (deficiency, or def.) of chromosomal material. Although changes from wild type to *Bar* always involved placing bands 16A1–2 adjacent to unfamiliar chromosomal bands, the reversions from *Bar* to wild type show that not every unfamiliar band lying next to 16A1–2 results in the *Bar* phenotype. ("No change" means "no cytologically visible change.") (After Sutton, 1943, by permission of *Genetics*.)

unequal crossing-over might do (Figure 6-3). Another, 263-20, neatly removed both duplicated segments. Both 263-46 and 263-20 proved to be lethal to hemizygous males and to homozygous females. (Revertant wild-type chromosomes that arise through unequal crossing-over are not lethal, of course.)

The remainder of the revertants, if cytological changes are detectable at all, involve deficiencies and inversions of chromosomal material. Note 257-6: bands 16A1–2 are placed adjacent to 15F1–2 (not 15F7, the usual neighbor), but a wild-type eye ensues. Similarly, inversion 263-5 places 16A1–2 adjacent to 17A5–6 and another inversion (rev. 2) places 16A1–2 next to bands 3F7–8 (inversion to 4A). Clearly, not every alien band that lies adjacent to 16A1–2 causes the *Bar* phenotype. The number of cases analyzed is small, but, on the basis of the available data, one might say that there is about a 50:50 chance that an unfamiliar chromosomal band lying adjacent to 16A1–2 will lead to *Bar* eyes; in half of such cases the eyes remain normal.

The third of Sutton's experiments (Figure 6-7) involved the reversion of *Double-Bar* and *Infra-Bar* to wild type. Once again, as in the reversion of *Bar,* one case (263-43) involves a virtually surgical removal of two of the three copies of region 16A1–7; only one copy remains. The removal was not accomplished without concomitant harm, however; 263-43 is lethal to males and to homozygous females.

There is no need to perform an extended analysis of Sutton's data. One might note, however, that, as in the case of *Bar* reversions, one wild-type revertant chromosome (257-9) removes the entire *Bar* region—and more. The 257-9 deficiency is, of course, lethal to males and homozygous females. Both it and the previous example (263-20 of Figure 6-6) result in wild-type rather than kidney-shaped eyes in heterozygous *Bar* females.

A missing portion of chromatin cannot produce products that correspond to the missing genes. However, the missing *Bar* region acts as a wild-type revertant. Thus, one is justified in concluding that the wild-type eye phenotype requires that chromosomal bands 16A1–2 *not act* at some moment during the embryonic formation of what will become the adult *Drosophila* head. The normal bordering bands in some manner have the ability to keep these bands from acting at the appropriate moment; about half of the bands in the salivary chromosomes represent chromosomal regions that possess this same ability. The other half do not, and they are the ones that, when moved to the neighborhood of 16A1–2, on either side, lead to the encroachment of normal head tissue into a region normally reserved for the fly's compound eye.

The analysis that Sutton carried out at the cytological level compares with that performed by Wright (p. 88) in his analysis of coat color in guinea pigs. Both, by careful reasoning and painstaking

B to + →

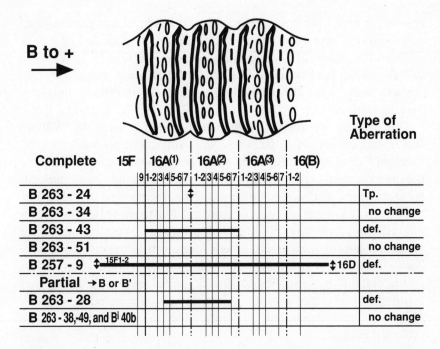

Type of Aberration

Complete	15F	16A(1)	16A(2)	16A(3)	16(B)	
		9|1-2|3|4|5-6|7	1-2|3|4|5-6|7	1-2|3|4|5-6|7	1-2	
B 263 - 24			↕			Tp.
B 263 - 34						no change
B 263 - 43		▬▬▬▬				def.
B 263 - 51						no change
B 257 - 9 ↕ 15F1-2	▬▬▬▬▬▬▬▬▬▬▬▬▬▬				↕16D	def.
Partial → B or B′						
B 263 - 28			▬▬▬▬			def.
B 263 - 38,-49, and Bⁱ 40b						no change

Figure 6-7. The *Bar* region triplication of *Ultra-* (or *Double-*) *Bar* and the aberrations that lead to the reversion of the mutant phenotype to wild type. Notice that here, as in Figure 6-6, deletions of the chromosomal segment 16A1–7 are often extremely precise; thus, they resemble a crossover-type event rather than one of random breakage under irradiation. Note, too, that deletions of the *Bar* region, although lethal in homozygous condition, behave as wild-type chromosomes in females that are heterozygous for *Bar* (def/B). (After Sutton, 1943, by permission of *Genetics*.)

observation reached conclusions that were not earlier evident to colleagues and fellow workers. Both lacked the tools and methods that are available to modern, molecular geneticists. An often overlooked fact may be emphasized here: *Had these early workers lived today and possessed today's techniques, they would be among today's leading geneticists.*

Variegated position effect

Earlier in this chapter (see Figure 6-4), we touched on variegated position effect as one of two recognized types of position effect, but then we turned our attention to the stable *Bar*-eye phenotype. The reason for returning now to the variegated type is that genetics and geneticists thrive on variation, on differences. The science of genetics from the outset has concentrated on the question, Why do these two (or more) individuals differ? As Haldane once pointed out, to ask why individuals are the same is a different, and much more difficult, question.

The present section will rely heavily upon reviews by William Baker and, earlier, by E. B. Lewis. The comments encountered here, then, will deal with studies on gene action—via position effect—through the mid-1970s. More recent studies, those involving molecular techniques, will be dealt with in the final section of this chapter.

What was known of gene action that led to variegated position effects? And how was this knowledge gained? These matters were restricted almost exclusively to one organism, *Drosophila melanogaster.* The experimental procedures that were developed in an effort to understand the genetics of this organism were also those that resulted in variegated position effect, and an understanding of these position effects demanded still more experimental procedures. The cycle fed on itself.

The "variegation" of variegated position effect arises through the suppression of normal gene activity in certain somatic cells. Thus, a red-eyed *Drosophila* may possess patches of white facets because the normal allele (w^+) at the *white* (w) locus has been suppressed and is not functioning properly. The suppression is precisely that: it is not a matter of mutation. One w^+ allele can be substituted for another in the genome that gives rise to variegated position effect. Each w^+ that is inserted into the proper milieu produces a variegated phenotype; upon recovery (removal) by recombination, it proves to be a normal wild-type allele once more.

Two technical points should be covered before plunging into a discussion of variegated position effect. Both deal with *D. melanogaster* but, because that is the species for which most data exist, some technical information is needed. The first deals with chromosomes and their physical rearrangements. The chromosomes that can be observed most readily are the giant ones in the salivary gland cell nuclei. The discussion of the *Bar* duplication centered on the information yielded by these chromosomes. One must remember, however, that eyes are not formed in salivary glands. The effects of the *Bar* duplication take place among head capsule cells and their

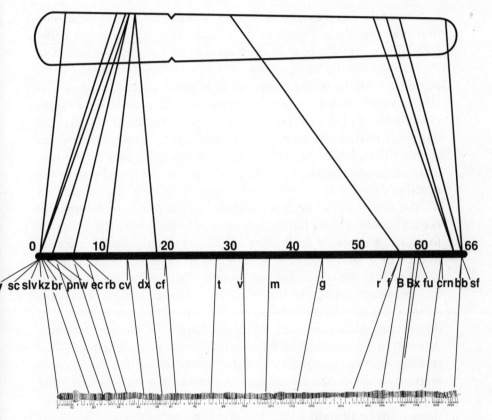

Figure 6-8. Comparison of the genetic map of the X chromosome of *Drosophila melanogaster* (center) with the metaphase chromosome of the mitotic cycle (top) and the giant chromosome of a salivary gland cell (bottom). Although the *order* of the gene loci (indicated by symbols only) remains the same, relative distances can vary considerably. The numbers 0–66 in the center of the diagram are map distances.

embryonic precursor cells. The chromosomes of these cells need not reflect the salivary chromosomes in every detail. Figure 6-8 presents the cytological and genetic maps of sex-linked genes occupying the X chromosome of *D. melanogaster*. One gains an immediate impression of stretching and shrinking: genetic map distances do not necessarily reflect physical distances directly because crossovers (recombination) may occur more often at one physical site on the chromosome than on another. The *order* of genes on the three maps is, of course, the same.

A second point is that some 20% of the metaphase X chromosome that lies near the centromere is reduced to a few dark but badly formed bands at the base of the salivary gland X chromosome. The explanation for the "disappearance" of this amount of mitotic chromosomal material is that it represents *heterochromatin* and that heterochromatin does not undergo repeated replication in the salivary gland nucleus. (The heterochromatic centromeric regions of all chromosomes coalesce to form an ill-defined body called the chromocenter.) Thus, while the *euchromatic* portion undergoes a 1000-fold increase in amount, the heterchromatic portion of the chromosome remains virtually, if not absolutely, constant in quantity. In addition to the centromeric heterochromatin, *interstitial* heterochromatin occurs here and there farther out on the chromosome. This also fails to undergo repeated replication in salivary gland cells. One consequence of a failure of heterochromatin to replicate in the salivary gland cells is the appearance of seeming "hot spots" for chromosomal breakage, especially following X-radiation. The chromocenter seems especially prone to breakage (see Figure 6-5, for example), as do certain smaller regions along the length of the chromosome. Actually, the probability of being broken for a given length of DNA remains constant; all the breaks that occur within the heterochromatin only appear to accumulate in isolated hot spots because the heterochromatic portions, having failed to undergo repeated replications, form small spots within the giant chromosomes.

A second technical point should be covered because the term *variegated* implies that gene action in different cells may differ. At some point during development, a gene with an obvious phenotypic effect (e.g., eye color) is programmed to function or not to function. Cells that descend from the one in which the decision is made form a clone—one that, in the eye, may eventually have red pigment or none.

Many cell lineages have been worked out in *Drosophila,* but not all by any means. (Workers studying the small nematode *Caenorhabditis elegans* have, in their case, traced the source of every one of the nearly 1000 cells that form an individual worm.) One method for studying cell lineages is illustrated in Figure 6-9. Eggs and young larvae that were heterozygous for the sex-linked mutant *forked* (*f*) bristle (and, hence, female because they possess two X chromosomes) were exposed to low levels of X-radiation. Some were exposed at age 12 hours or less, others at 24, 48, and 72 hours.

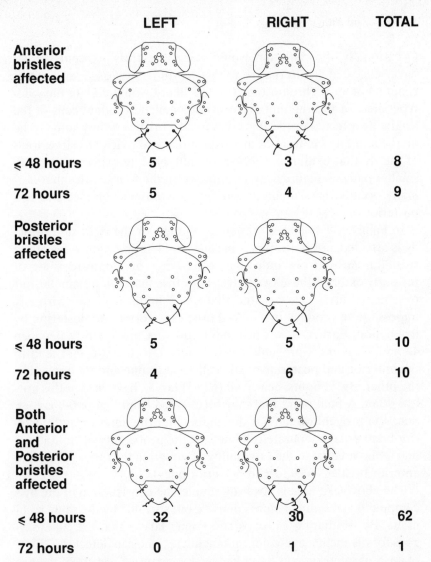

	LEFT	RIGHT	TOTAL
Anterior bristles affected			
≤ 48 hours	5	3	8
72 hours	5	4	9
Posterior bristles affected			
≤ 48 hours	5	5	10
72 hours	4	6	10
Both Anterior and Posterior bristles affected			
≤ 48 hours	32	30	62
72 hours	0	1	1

Figure 6-9. The distribution of *forked* bristles on the scutellum of a total of 100 flies (*Drosophila melanogaster*) that were exposed early in development to X-radiation. Some were exposed to radiation 12, 24, or 48 hours after being deposited as eggs (≤48 hours); others were irradiated 72 hours after eggs were laid. Note that, very often, an exposure to radiation resulted in forked bristles at both the anterior and posterior bristle scutellar sites when exposure occurred 48 hours or earlier; at 72 hours, only one instance occurred when both the anterior and posterior bristles were affected. Both the left and the right sides of the scutellum arise as single cells lying on opposite sides of the embryo; by 72 hours after egg laying, the common precursor cell for the anterior and posterior bristles has divided so that the cellular lineages leading to these bristles become independent entities. (Wallace, unpublished data.)

Occasionally, the radiation would cause the "loss" (recent studies have suggested an alternative to *loss;* it will be discussed later in this chapter) of an X chromosome (or, less likely, would mutate the wild-type allele of f to the mutant form) of a cell; descendant cells of the originally affected one will possess forked bristles if they come to lie on the adult fly's exterior. Early exposures tend to lead to large areas of the fly that exhibit the *forked* phenotype; late exposures lead to smaller patches. Figure 6-9 concentrates on the four bristles that normally occupy the scutellum: two anterior scutellar bristles and two posterior ones.

In Figure 6-9, one sees that either the left or the right side of the fly is affected, not both. This is because the left and right sides of the scutellum arise from individual cells lying on opposite sides of the early embryo; these two halves later fuse just as, for example, our own upper lip. Next, we note that at 72 hours it proves virtually impossible to produce two forked bristles, anterior and posterior, by irradiation. Earlier, it was possible to do so. Consequently, between 48 and 72 hours, a progenitor cell divides into the two that lead to the anterior and posterior bristle cells, one to one and the second to the other. By 72 hours, nearly all (62/63) larvae have passed this critical point. A final point that may be inferred from Figure 6-9 is that, once the progenitor cell has divided so that the anterior and posterior bristles have separate cell lineages, the probability of "mutating" one is about equal to the probability of mutating the other (17 forked anterior bristles to 20 forked posterior ones).

The above example is absurdly simple; the construction of the fly's compound eye is much more complex, but solvable by the same technique. By irradiating embryos and young larvae that are heterozygous for distinctive eye-color mutant alleles, one can detect the "loss" of an X chromosome in a cell lineage by means of sectors of the eye that exhibit an appropriate color. The irradiation of a w^+/w heterozygous female larva will generate *white* sectors following the "loss" of the w^+-bearing X chromosome. Because of the numerous alleles at the *white* locus, many other combinations of alleles are both suitable and available for study.

A consistent pattern of sectoring (Figure 6-10A), in which the individual sectors can be labeled I–VIII as in the figure, *suggests* that the eye primordium may possess eight cells, each of which is responsible for one of the eight sectors in the ventral position of the compound

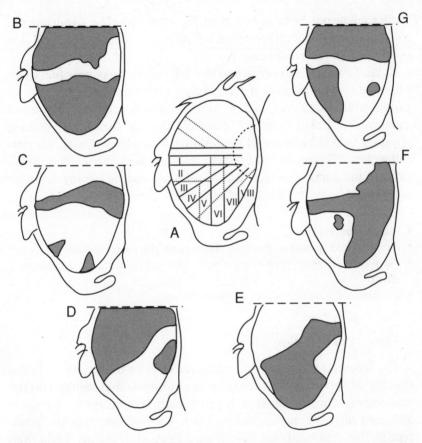

Figure 6-10. Patterns of mottled areas of the compound eye (*Drosophila melanogaster*) that arise from variegated position effect. These are to be compared with the center diagram (A), which illustrates the patterns of mutant tissue that ensue from the exposure of embryos and young larvae to X-radiation. The similarity of these patterns suggests that the suppression of normal genes by nearby heterochromatin occurs as a distinct event, in a particular cell, and that the descendants of that cell lack that gene's normal activity. (After Baker, 1963, by permission of *American Zoologist* and W. K. Baker.)

eye. With respect to variegated position effect, it is sufficient to note that the variously colored sectors of the eye that result from the inactivation of a wild-type allele fall into the same patterns as do the sectors that arise by an early "loss" of an X chromosome. Indeed, if we recall the kidney-shaped eye of heterozygous *Bar* (i.e., +/*B*) females, it might be that the facetless notch on the eye's anterior border corresponds to sector II in Figure 6-10A. The experimental results *suggest* this interpretation, but subsequent analyses reveal that matters are more complicated.

Baker (1968, p. 160) remarked that his reader "may now feel overwhelmed by the myriads of facts known about position-effect variegation which have yet to fall within a simple, rational biological framework." Baker's remark parallels that of the theory-making physicist-turned-biologist (Leo Szilard) who pleaded with his new colleagues, "Please do not confuse me with your facts!" The observations and facts to which Baker referred can be arranged under several headings and discussed in serial fashion.

- the *cis-trans* relation
- the polarized spread of position effect along the chromosome
- the Y-chromosome suppression of most instances of variegated position effect
- the parental modification of position effect

The *cis-trans* relation

The terms *cis* and *trans,* one might recall, refer to the physical relationship of two pairs of genes (or other chromosomal regions) on the homologous chromosomes that carry them. If we refer to the genes and their differing alleles, *A-a* and *B-b,* we might say that the genotype *AB/ab* represents the *cis-* arrangement, while *Ab/aB* might then be referred to as the *trans-* arrangement. This usage corresponds to the much older terms *coupling* and *repulsion.* Having specified which combination interests us—*AB, Ab, aB,* or *ab*—we can say that that combination is in the *cis-* arrangement if the two are on the same chromosome.

For variegated position effect, the break point (*bp*) that lies within the heterochromatic portion of a chromosome is one item of interest; *bp*$^+$, then, signifies the unbroken region of the homologous chromosome. The other item of interest is a gene for which the normal allele (*g*$^+$) and a hypomorphic mutant allele (*g*) are known. Letting *R* (rearrangement) designate the chromosomal rearrangement that possesses the break point (*bp*), we can identify two genotypes: (1) *R* (*g*$^+$ *bp*)/*g bp*$^+$ and (2) *R* (*g bp*)/*g*$^+$ *bp*$^+$.

Genotype 1 exhibits variegated position effect. The wild-type allele

(g^+), through its association with the heterochromatic break point (*bp*), has its normal activity suppressed in certain cell lineages. Genotype 2 in which g^+ is not associated with heterochromatin, leads to a nonvariegated, wild-type phenotype. The "inactive" mutant allele, *g*, may or may not be subject to additional suppression because of its proximity to heterochromatin (*bp*); such suppression would not be revealed phenotypically.

The polarized spread of position effect

The suppression of normal alleles through the action of nearby heterochromatin is a spreading one, one that spreads in a polarized manner along the *cis* chromosome (see Figure 6-4). An elegant demonstration of this polarized spreading effect is illustrated in Figure 6-11. The euchromatic segment containing the two loci, *white* and *split*, has been involved in two chromosomal rearrangements. At the left, the two loci have been attached to the heterochromatin of chromosome 4; at the right, the two loci (and the segment of euchromatin within which they are located) have been inserted within the heterochromatin of the left arm of the V-shaped chromosome 3. Examination of the eye diagrammed on the left reveals that all facets of the eye that are white are also rough (an effect of mutations at the *split* locus). The suppression "flows" from the heterochromatin to *split* and, at times, to *white* as well; if it reaches *white*, it has always reached the physically closer *split*. On the right, *white* facets may or may not be rough (*split*), and rough facets may or may not be white. In this case, the "spread" of gene suppression may come from either end of the inserted euchromatic segment, sometimes reaching *white* from the right end (but not *split*), sometimes reaching split from the left (but not *white*), and sometimes reaching both loci from either or both sides.

Y-chromosome suppression of position effect

Another detail, unexplained but real, involves the Y chromosome, which in *D. melanogaster* is almost entirely heterochromatic. (In salivary gland preparations, the Y chromosome is represented by a small, ill-defined mass of chromatin that is associated with the heterochromatic centromeric regions of the other chromosomes, the chromo-

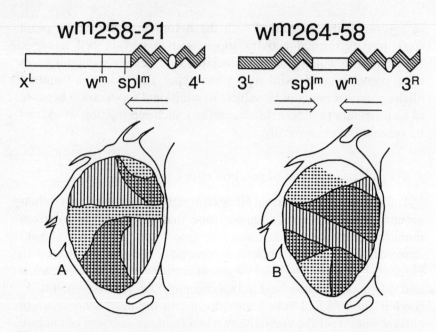

Figure 6-11. Diagrams illustrating (top) two chromosomal aberrations. At the left, the tip of the X chromosome carrying the wild-type alleles at the *white* and *split* loci has been translocated to chromosome 4. At the right, the segment of the X chromosome carrying the same two loci has been *inserted* within the heterochromatin of chromosome 3. The lower diagrams reveal, on the left, that every white facet of the compound eye is also rough (the effect of mutant alleles at the *split* locus) and, on the right, that white facets may be either rough or smooth. The open arrows at the top indicate the spreading of position effect into the euchromatic segment. Stippling indicates "rough" areas, vertical hatching indicates pigment, and absence of hatching indicates white areas. This figure may be compared with Figure 6-4. (After Baker, 1963, by permission of *American Zoologist* and W. K. Baker.)

center.) The presence of a Y chromosome reduces either the frequency of gene suppression by adjacent heterochromatin or the probability that such suppression will occur. The latter, by its delaying action, may cause the patches of variegated tissues to be small or even to be missing altogether. The effect of the Y chromosome is readily detected by comparing the phenotypes of XO and XY males, or XX and XXY females.

The ability of the Y chromosome to suppress the inactivation of normal alleles by nearby heterochromatin (position effect) has been used in demonstrating that all genes—not merely those genes that affect the visible phenotype of the fly—are subject to this type of inactivation. This demonstration stems from the well-known ability of X-radiation to induce gene mutations, many (if not most) of which are recessive lethals. Through the genetic manipulations that special strains of *D. melanogaster* allow, it is possible to test the same irradiated X chromosome in two types of males: XO (males lacking a Y chromosome) and XY (normal males). Many irradiated X chromosomes (20%) prove to be lethal in XO males but not in the presence of the normal male's Y chromosome (Lindsley et al., 1960). When examined cytologically, all of the Y-chromosome rescuable lethals proved to be associated with chromosomal aberrations in which euchromatic portions of the X chromosome were now joined to heterochromatin. The breaks in the X chromosome were scattered throughout the length of that chromosome; the constant feature was that the break involved heterochromatin. Thus, genes that, when mutated, give rise to recessive lethals appear to have a similar lethal effect when inactivated because of (variegated) position effect. So XO males that carry these rearrangements die. However, the position effect is itself suppressed by the presence of a Y chromosome; thus, in normal males, the wild-type alleles of these many potentially lethal loci are not inactivated and the XY males live.

Virtually everything that has been said about variegated position effect has exceptions. Because position effect, itself, is poorly understood, nothing is to be gained by an extensive account of exceptional instances. An exception that can be mentioned specifically, however, is that of genes that are normally located in heterochromatin. When such genes are relocated into euchromatin that is far removed from heterochromatin, they also exhibit variegated position effect. They exhibit variegated position effect as well when relocated near centromeres, even though, there, they remain embedded in heterochromatin. The exceptions and the complexities are such that they appear to rule out simple explanations; this appearance will probably prove wrong when the molecular basis of gene suppression is understood.

The parental modification of position effect

As if the observations, rules, and exceptions to rules that have emerged from the observations and thoughts of dozens of drosophilists have not proved to be sufficiently puzzling, one more

Table 6-1. Data showing the effect of parental genomes on the expression of variegation. (After Noujdin, 1944.)

Mother	Father	Offspring	Percentage variegated
sc^8/sc^8	$+/Y$	$sc^8/+$	4.8
$sc^8/+$	$+/Y$	$sc^8/+$	40.0
sc^8/sc^8	sc^8/Y	sc^8/sc^8	13.9
$sc^8/+$	sc^8/Y	sc^8/sc^8	40.6

complicating factor can be added: the degree of variegation in an individual is a function of the genotype of the parent. Crosses that produce offspring of identical genotypes need not produce the same degree of variegation in those offspring (Table 6-1).

In studying variegation at two loci, *achaete* (*ac*) and *yellow* (*y*), N. I. Noujdin, a Soviet geneticist, tabulated the proportion of progeny flies that exhibited any variegation, whatsoever. The results are shown in Table 6-1. Flies heterozygous for the *scute*-8 (sc^8) inversion, an inversion that rotates virtually the entire X chromosome because of one break near the tip of the chromosome and the other in the chromocenter (basal heterochromatin) exhibit variegation of *achaete*. (A prominent effect of mutants at this locus is to "remove" the large posterior dorsal-central bristles from the flies' thorax; they have other phenotypic effects as well.) If the mother is homozygous sc^8/sc^8, only 5% of her progeny lack these bristles; if she is heterozygous $sc^8/+$, the proportion of offspring lacking bristles is 40%. Fragments of the X chromosome and of the Y chromosome are also capable of exerting a parental influence on the proportion of variegated offspring even though the offspring are of but one genotype (see Baker, 1968).

The molecular biology of heterochromatin

The terms *heterochromatin* and *euchromatin* have been used throughout this chapter as they have been used throughout much of the history of genetics. During the mitotic cycle, certain regions of chromosomes take up cytological stains at times when other regions either fail to do so or do so to a much lesser degree. The "unexpected" heavily staining regions (heterochromatin) often surround the chromosomes' centromeres—often extending 20%–30% of the length of the chromosome arm.

Only during the early 1930s have geneticists realized that the large, banded nuclear structures of dipteran salivary gland cells are chromosomes that correspond to the much smaller ones that are found in dividing somatic cells. The giant chromosomes allowed, of course, for the precise determination of gene loci that has been cited so frequently throughout this chapter. Physically, there is considerable difference—besides the matter of size—between salivary and mitotic chromosomes: The large areas of heterochromatin, including the entire Y chromosome, are underrepresented in salivary nuclei. The huge volume of heterochromatin in other somatic cells is reduced to a comparatively small mass—the chromocenter—in salivary gland nuclei. Heterochromatin either is not replicated in these nuclei or is replicated to a much smaller extent. (Euchromatic chromosome arms in these cells consist of 1000 or more parallel strands, the result of ten chromosomal replications.)

Today, heterochromatin is recognized as consisting largely of DNA whose composition is simple and repetitive. In *Drosophila melanogaster*, for example, heterochromatin constitutes about one quarter of the entire genome. When the DNA of *D. melanogaster* is subjected to centrifugation in a cesium chloride gradient, six distinct bands, or "peaks," are observed. In situ hybridization tests show that these represent the centromeric regions of all chromosomes as well as virtually all the fourth and Y chromosomes. The composition of one peak, for example, consists of 1 or 1.5 million copies of ... AATAT ... or ... AATATAT.... Another peak represents 1 million copies of a 10-base-pair unit. The remaining peaks are also composed largely of simple (and some not so simple) repetitive sequences. This feature, one can see, sets heterochromatin quite apart from euchromatin, which, of necessity, must contain large numbers of "single-copy" DNA—that is, of DNA that specifies the amino acid sequences of thousands of different proteins. In addition to these protein-specifying regions, however, euchromatin possesses numerous gene control regions that determine when and in what cells the proteins will be made.

Ordinarily, biologists speak jointly of *structure* and *function*. Here we have provided a cursory view of the structure of heterochromatin. What about its function? There is no good answer to that question at the moment. Speculations abound, simply because there are many persons who are willing to speculate. So many proposed functions have proved to be erroneous, that several persons have proposed that heterochromatin really has no function. Reference to Bernard John and George L. Gabor Miklos's book *The Eukaryote Genome in Development and Evolution* (1988) seems to be an appropriate ending for this chapter; readers truly interested in this topic will find many items of interest in that text.

7 Using the Environment as a Research Tool

From the outset, geneticists recognized that the appearance of an organism is determined in part by its heredity and in part by the environment in which it lives and develops. Mendel's first admonition for the proper selection of experimental plants was that they "possess *constant* differentiating characters." Stated differently, the characters were not to vary excessively in response to different growing conditions. Or, alternatively, under normal conditions, these characters must not vary.

Students of elementary genetics, even of general biology, learn of the Himalayan rabbit, which is characterized by its black nose, ear tips, feet, and tail. The pigmented areas are located on the animal's extremities, so there is reason to believe that pigmentation occurs at a relatively low temperature. Proof of this hypothesis has become "common knowledge": Himalayan rabbits raised at low temperature are black, those raised at high temperature are white, and patches of white fur that are removed and regrown while the animal is kept at low temperature are pigmented. By "common knowledge" is meant that literature citations no longer accompany descriptions of these experiments. Actually, they were initially carried out by two German geneticists, W. Engelsmeier and Rolf Danneel, during the mid-1930s. Knowing what one knows about melanin formation (Chapter 2), one may suggest that the Himalayan rabbit possesses a temperature-sensitive enzyme (probably tyrosinase) that functions at low temperature but not at a temperature that equals or exceeds the rabbit's

Figure 7-1. An ear of corn borne by a mutant ("weak sun red") plant. On this plant, husks that grow at night remain green (pale in figure), whereas husks that grow during daylight hours become red (dark in the figure). Cells that develop in the dark lack the ability to synthesize anthocyanin, a red or purple plant pigment. (Redrawn from Emerson, 1921.)

normal body temperature. Temperature-sensitive mutants will recur later in this chapter; they constitute one of the most powerful of all genetic tools.

Corn farmers from ancient times must have been aware of certain strains that develop purple (anthocyanin) pigmentation only in those tissues of the plant that develop in the presence of sunlight. Not to notice the alternating green and purple stripes on the husks covering the ears of this type of corn is nearly impossible. The elongation of the ear and its husk that occurs at night is marked by green tissue that is unable to produce anthocyanin even upon later exposure to sunlight; the tissue that forms during the day becomes pigmented as it emerges (Figure 7-1).

Finally, the gross change in phenotype that occurs in plants that are adapted to both terrestrial and aquatic life may be mentioned (Figure 7-2). The arrowleaf (*Sagittaria saggitifolia*) is so named because of the

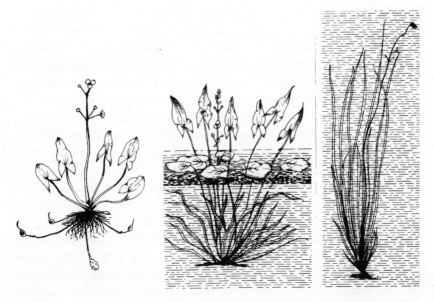

Figure 7-2. The adaptive response of the arrowleaf (*Sagittaria sagittifolia*) to growth in and out of water. A plant that starts growth in water only to emerge above the surface develops a morphology appropriate for each environment. (From Wallace and Srb, 1961, by permission of Prentice-Hall.)

shape of its leaf *as it grows on land* or on and above the surface of a pond. These leaves have the shape of a stereotypical arrowhead; each is supported at the base of a rather rigid stem. Such leaves and stems are excellent for gaining exposure to sunlight in the air but would be ill-designed for existence under water, where they might be subjected to water currents. Arrowleaf plants that grow submerged (either the entire plant or the submerged portion of a plant that has broken through the water's surface) have flexible, ribbonlike leaves that are buoyed by the water (hence, no need for stems) and undulate with the currents. These leaves, lacking a cuticle, absorb much of their nutrients from the water; the plants' underwater root systems are poorly developed. Although it is relatively simple both to describe and to account for the differing aquatic and terrestrial forms, virtually nothing is known of the genetic mechanisms that control these diverging developments.

Temperature

The study of gene action by the manipulation of the environment, if pursued rigorously, would include the effects of such highly specific acts as overwhelming the normal human body's ability to metabolize homogentisic acid by ingesting abnormally large quantities of the acid (Chapter 2). Or, even more detailed, the provision of particular vitamins or amino acids to mutant strains of *Neurospora* or bacteria. That, however, is not our intention here. In this chapter, we shall discuss (1) attempts to learn of genes and how they perform their functions by comparing the results obtained by raising genetically similar (as nearly identical as possible) organisms under different environmental regimes, (2) attempts to pinpoint the critical times at which different environments have their effects, (3) the recognition that many environmental manipulations lead to phenotypic consequences resembling those of known mutant genes (phenocopies), and (4) the use of temperature-sensitive mutants in unraveling the time of gene action and in identifying essential genes and gene products. Much of the work, merely because it requires an adequate control of the prevailing environment, will concern experiments involving temperature that were carried out in the laboratory. Furthermore, because an understanding of gene action is the goal of these studies, *Drosophila melanogaster,* the workhorse of early geneticists, will frequently be involved.

Metabolic rate

An obvious approach by early geneticists, in their attempt to understand gene action, was to use a favorite approach of physiologists: determine the effect of temperature on reaction rates. For many of these physiological studies, including those on developmental rates, the ultimate goal was to determine the increase in rate of reaction for each 10°C increase (Q_{10}) in temperature. Acknowledging, of course, that both abnormally high and abnormally low temperatures may cause vital reactions to cease altogether (including death of the treated organism), Q_{10} values of about 2 are commonly observed. In the case of *Drosophila melanogaster,* the time from egg laying until pupation is about 116 hours at 25°C, but only 100 hours at 30°. The pupal stages require about 112 hours at 25° and 80 hours at 30°. The

Q_{10} for development within the pupal stage is almost exactly 2 [112/80 = 1.4 = Q_5; $(1.4)^2 = 1.96 = Q_{10}$].

Genetic assimilation

From such physiologically oriented studies evolved others that dealt with the effect that temperature might have on mutant as well as on wild-type flies. For example, flies grown at low temperature are larger than those grown at higher ones, and *Curly* winged flies grown at low temperatures tend to have noncurly (straight) wings. Both of these examples deserve a brief additional comment. *Drosophila* flies of various species tend to be larger in the cooler portions of their geographic ranges than in the warmer regions; when the strains from different localities are grown in the laboratory at the same temperature this difference in body size persists. The difference, which is a *physiological* difference demonstrable by raising a single strain of flies at high and low temperatures, becomes a *genetic* difference that contrasts warm and cool populations or races of flies. Furthermore, the switch from a purely physiological temperature effect to a genetically based one has been demonstrated in laboratory populations of *Drosophila pseudoobscura* that were maintained at low (16°C) and high (25° and 27°) temperatures for six years. After an additional six years (12 years total at 16°, 25°, and 27°), genetic differences in size could be demonstrated for populations kept at all three temperatures (16° versus 25° versus 27°) (Anderson, 1973). The changeover from physiological to genetic control of the phenotype can be explained by genetic changes wrought by natural selection; the process is known by the term *genetic assimilation* (Waddington, 1953).

Curly-winged flies (Figure 7-3) present a case seemingly less complicated than the genetic assimilation of body size described above. The description of this mutant reads (Lindsey and Grell, 1967, p. 65): "Wings curled upwards; rarely overlaps wild type at 25°, but frequently overlaps at 19°C." That statement is correct, except that by selecting as parents only those flies with curled wings at 19°C one can rather quickly establish a strain of flies whose wings are clearly distinguishable from wild type at that lower temperature. The expression of the gene, *Curly,* is governed not only by temperature but also by the background (modifier) genotype. Numerous mutant genes listed in the catalogue *Genetic Variations of* Drosophila

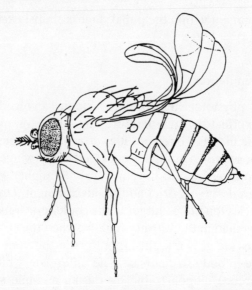

Figure 7-3. The *Curly* (wing) mutant of *Drosophila melanogaster*. As a rule, the expression of this mutant (tightness of curl) is enhanced at 25°C or above; at lower temperatures, the wing may appear to be wild type (noncurled). Artificial selection for curled wings at low temperature yields flies that express the mutant phenotype at low temperatures.

melanogaster (Lindsley and Grell, 1967) are said to overlap wild type as the result of temperature effects, culture conditions, or unexplained "environmental" effects.

Temperature-sensitive periods

Because temperature is an easily controlled environmental agent in both degree and duration, a search for temperature-sensitive periods during development became an obvious task for early drosophilists. An example is provided by an extended study carried out in the 1930s on flies carrying various mutant "alleles" at the *Bar* locus in *D. melanogaster*. In this case, the flies were "*Bar-Infrabar*"; these flies carried the *Bar* duplication, but one of the duplicated chromosomal segments, as the result of a submicroscopic alteration, had a less severe effect on facet number (so-called *Infrabar*) than did the original *Bar* duplication (see p. 106).

When homozygous *Bar-Infrabar* flies are raised at different (but

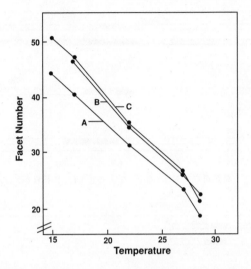

Figure 7-4. The effect of temperature on the number of facets in female (A and B) and male (C) *Bar-Infrabar Drosophila melanogaster*. At $28^{1}/_{2}$°C the flies' eyes are scarcely half as large as they are when raised at 15°C. Curve B was obtained 17 months after curve A, thus demonstrating the consistent pattern of these observations. (Redrawn from Luce, 1935, by permission of the *Journal of Experimental Zoology*.)

constant) temperatures, the mean number of facets in their compound eyes differ. At 15°C, males have an average of 44 facets and females, 51; at 17° these comparable numbers are 40 and 47; at 22°, 31 and 35; at 27°, 23 and 27; and at 28.5°, 19 and 21 (see Figure 7-4). Interestingly, all *Bar* mutants have fewer facets (smaller eyes) when raised at high temperatures *except* homozygous *Infrabar* and homozygous *double-Infrabar* (B^iB^i/B^iB^i); the eyes of these flies become larger when raised at increasingly high temperatures. Such puzzling reversals of the phenotypic effect of temperature differences (also true, for example, of different mutant alleles at the *vestigial* locus, a locus whose genes affect wing size) offer to some persons a hope for understanding the underlying causative events. On the other hand, they discourage other, more faint-hearted persons who view matters as being hopelessly complex.

The experiments on the effects of temperature on facet number were extended to determine whether these effects are spread throughout the larval life span or limited to specific periods during larval life when a given temperature exerts its effect. The latter

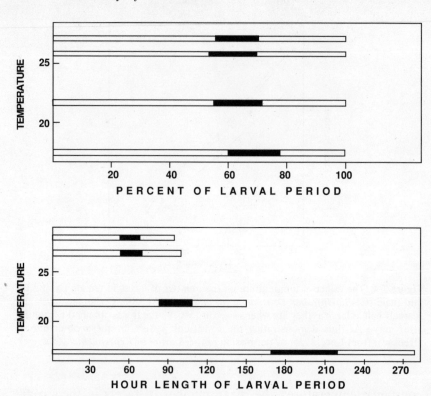

Figure 7-5. A diagram revealing the percentage and duration (in hours) of the effective period for temperature (°C) in affecting the number of facets in eyes of *Bar-Infrabar Drosophila melanogaster*. The solid portion of a bar indicates the temperature-effective period. The lower half of the figure shows time in hours; the upper half transforms larval life into percentages (100% from egg to pupation). (After Luce, 1935, by permission of the *Journal of Experimental Zoology*.)

proved to be the case (see Figure 7-5). The experiment proceeded by starting the development of homozygous *Bar-Infrabar* individuals at one constant temperature and transferring them at definite times to a second temperature. The results are shown in Figure 7-5 and Table 7-1. Note that the larval period lasts considerably longer at low than at high temperatures; this is apparent in the lower half of Figure 7-5. At 17°C, the mutant flies spend nearly 12 days as larvae before pupating (the lowermost bar on Figure 7-5). After these differences in developmental time have been properly adjusted, one sees a consistent pattern. When about 60% of the larval life has passed, there

Table 7-1. Data revealing that the number of facets in the eyes of *Bar-Infrabar* adult *Drosophila melanogaster* that have been exposed to a given temperature midway during their development is the same as if they had undergone their entire development at that temperature. (After Luce, 1935.)

Entire development		Temperature manipulation		Temperature manipulation	
Temperature (°C)	Mean no. of facets	Time spent at each temperature (hours)	Mean no. of facets	Time spent at each temperature (hours)	Mean no. of facets
17	46.1	27° 54 17° 52 27° remainder	44.1	27° 58 17° 50 27° remainder	43.6
22	34.6	28½° 54 22° 23 28½° remainder	31.9	28½° 54 22° 24 28½° remainder	33.1
27	25.8	17° 172 27° 18 17° remainder	26.3	17° 174 27° 16 17° remainder	26.8
28½	22.7	22° 83 28½° 15 22° remainder	23.4	22° 84 28½° 13 22° remainder	25.4

comes a period during which the temperature that prevails determines the size of the adult eye of *Bar-Infrabar* individuals. Table 7-1 confirms the results shown in Figure 7-5. In each of the eight experimental tests, the number of facets of adult *Bar-Infrabar* individuals that spent the intermediate period of their larval life at a given temperature corresponds to that of flies that spent their entire lives at that temperature, regardless of the temperatures that bracket the intermediate period. Thus, high temperature that both precedes and subsequently follows an intermediate exposure to cold temperature does not lower the facet number of adult eyes. Inversely, cold temperature at both the start and end of larval life does not increase the number of facets for flies whose intermediate hours were spent at high temperature.

An observation often made by drosophilists is that the normal, or wild-type, phenotype is much more stable than that of mutant or otherwise abnormal flies. Mention has already been made of the many instances in which mutant phenotypes overlap wild type under various environments, including various temperatures. Intersexes, flies whose ratio of autosomes to X chromosomes is not 2:1 (male) or 1:1 (female), exhibit phenotypes ranging from nearly male to nearly female; in contrast, the two sexes are, of course, clearly male or clearly female in virtually every case. (One might infer—see Figure 5-5—that, as a rule, the capacity to produce an excess of enzymes and other proteins is constantly held in check by feedback mechanisms; the result is a stable phenotype. Variable phenotypes, such as those arising under stressful temperatures under this view, would arise when a critical compound fluctuates in amounts less than the normal ceiling, with the phenotype reflecting in each instance the amount actually present.)

Phenocopies

The wings of some insects, notably moths and butterflies, exhibit complex patterns that may vary from one geographic locality to another. A butterfly enthusiast, for example, would know that the Palestinian race of the swallowtail *Papilio machaon* differs from members of the same species living in Central Europe. Or that the tortoise shell, *Vanessa uritica,* comes in two varieties: *ichnusa* (southern) and *polaris* (northern). Perhaps it should not be unexpected that entomologists during the late nineteenth century, having learned that heat shocks delivered to butterfly pupae can alter the expected wing patterns of emerging adults, would have spent much time attempting to convert one pattern into a second one resembling another known variety. The simplest example is provided by the tortoise shell butterfly: exposure of pupae to heat generates patterns resembling the southern (European) form; treatment with cold generates patterns resembling the northern form.

The phenotypically more uniform *Drosophila* proved to be sensitive to heat shocks, as well. Richard Goldschmidt, a German—then American—physiological geneticist, found that heat shocks, when applied to otherwise wild-type flies, could result in abnormal phenotypes resembling any one of the hundreds of mutants known for this fly. Such alterations were called *phenocopies*. Although the altered adult may resemble one or the other of many known mutants, it is not a genetic mutant itself. In every known case, the phenocopy always produced wild-type progeny, as it should if its gametes had remained unchanged.

Knowledge that the phenotypes corresponding to those of known mutations could be obtained by an abrupt interference of normal development at specific times encouraged many early geneticists to believe that they could now undertake a thorough study of gene action. Goldschmidt (1949, p. 49) expressed their view most succinctly:

> The study of phenocopies and their relation to mutation shows clearly that every departure from the norm by mutation can also be accomplished as a phenocopy. I do not hesitate to draw from this the inverse conclusion that any departures from the norm produced by the action of phenocopic agents should also be obtainable as mutants, even in those cases in which no presently known mutants supply the phenotype resembling the phenocopy.

The analogy for normal development used by Goldschmidt was a system of railroad tracks with numerous switches; a railway car once set in motion by whatever cause is destined to travel the route that is determined by the setting of successive switches. (This analogy might be compared with the account of development expressed in modern molecular terms on p. 73.) The initial impulse may be

"normal," thus corresponding to initiating genetic signals, or "accidental," thus corresponding to phenocoptic agents (heat or cold
shocks, radiation, ether fumes, CO_2, and the like).

One of Goldschmidt's many studies involved feeding *Drosophila*
larvae on 0.06% sodium tetraborate, a diet that results in eyeless flies
(nearly 100% eyeless as lethality also approaches 100%; smaller proportions of "eyeless" flies with variable expression as the proportion
of surviving flies increases). A summary of many of his findings is
interesting: There are strains with low sensitivity and strains with high
sensitivity. Some strains produce no eyeless flies. The fourth chromosome (on which the mutant *eyeless* is located) of an extremely sensitive strain was sufficient, when transferred into the genome of a
refactory strain, to transform the latter into a sensitive strain.
Interstrain hybrids that exhibit hybrid vigor also exhibit resistance to
tetraborate. Abnormal phenocopies other than that of *eyeless* are
induced by borate—one in one strain, another in a second. Over-
all, Goldschmidt suggested that phenocopies result from bringing
into the open preexisting subthreshold effects: that is, the ability to
produce phenocopies has an underlying genetic basis. And, indeed,
that is what Waddington's "genetic assimilation" demonstrated
during the same era.

Perhaps the most carefully performed experiments on phenocopies
were those carried out by Roger Milkman during the late 1950s and
early 1960s. Following (but improving upon) Waddington's work on
the induction of *crossveinless* phenocopies by temperature shock,
Milkman (1962, 1967) exposed pupae of *D. melanogaster* to different
time-and-temperature combinations at various times following pupation. The results of two experiments are shown in Figure 7-6. Clearly,
the shorter exposure resulted in less-severe crossvein shortening than
did the longer exposure. Equally clear, however, is that there is a time
interval, 21–23 hours after pupation, during which the crossvein is less
liable to damage than either just earlier or just later.

These and complementary data led Milkman to conclude that a
complex metabolic pathway containing eight biochemical steps
including three reversible ones and one bypass loop could explain the
data. The difficulty with this analysis (the best and only approach of
that era), and with many others in the biological sciences, was that
the observations were far removed from their underlying causes. In
effect, there were more unknowns than equations. An attempt to

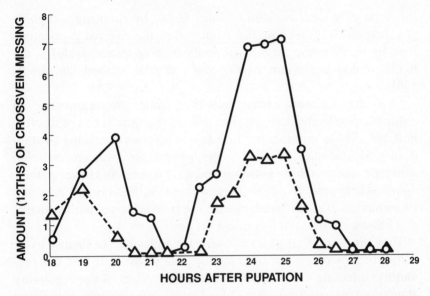

Figure 7-6. The effect of exposures of 30 (△) and 35 (○) minutes to 40.5°C at various times following pupation on the disruption of the posterior crossvein in the wings of *Drosophila melanogaster*. Note that there is a brief period between about 21 and 23 hours during which the flies' wings are more resistant to crossvein disruption than they are both earlier and later. (Adapted from the *Journal of General Physiology*, 1962, 45:777–799, by copyright permission of The Rockefeller University Press and by permission of Roger Milkman.)

understand one's observations when they are based on data such as those illustrated in Figure 7-6 corresponds to an attempt to recon-struct the "causative" definitions from an examination of a completed crossword puzzle. The definitions that are capable of leading to any finished puzzle are virtually innumerable.

Induction and repression

When exposed to any of a variety of agents (called inducers), cells will synthesize novel protein(s). Alternatively, some agents (called repressors) trigger the loss of a particular protein(s). Usually, agents of either sort trigger a rather narrow and specific response. In the late 1940s, three scientists at the Institut Pasteur in Paris, France—Jacques Monod, François Jacob, and André Lwoff—developed a model for this type of genetic regulation. They began by choosing a suitable genetic model; they chose the induction of the enzyme β-galactosidase by its substrate lactose (i.e., milk sugar, a disaccharide) in the human colon bacterium *Escherichia coli* (see Monod and Jacob, 1961).

First, they isolated mutant cells that, rather than expressing an inducible β-galactosidase, produced it all the time (i.e., constitutive mutants). These mutants fell into two categories according to standard genetic studies. Mutants in the first category were recessive to wild type, whereas those in the second were dominant. However, their dominance extended only to genes located on the same DNA strand (dominant in the *cis-* arrangement; see p. 65). Those characteristics, and others, led them to formulate the operon model.

They proposed that genes allowing the utilization of lactose by *E. coli* were subject to regulation by one gene whose only role was regulatory; it had no enzymatic function. The product of the regulatory gene (the *repressor*), which we now know to be a protein, could bind to a unique DNA sequence that is linked to the genes for lactose utilization. Once the product of the repressor had bound to that unique sequence (the *operator*), transcription of the β-galactosidase gene and others immediately linked to it was prevented. The operator gene was, like the repressor, novel. Further, it did not encode any protein product; it was merely a site for binding with the repressor. Mutations in the repressor-protein gene (constitutive in haploid cells) were recessive to wild type in diploids, because the normal repressor protein could diffuse through the cytoplasm and bind to any operator sequence and thus prevent transcription. By contrast, constitutive mutations that affected the operator sequence, thus leading to constitutivity, were dominant, because no repressor molecule could bind and prevent transcription. The final part of the model involved hypothesizing that lactose could bind to the repressor protein and prevent its binding to the operator.

There are interesting aspects to this story that contribute to our story of the study of gene action. First, the investigators felt that the basis for inducibility lay in the bacterium's genes; otherwise, they would not have attempted to isolate mutants that exhibited altered patterns of induction. Second, before the dominance and recessiveness of constitutive mutations could be determined, methods for constructing diploids of *E. coli* (albeit for small regions of bacterial chromosome) had to be developed. Thus, understanding the means by which *E. coli* could transfer DNA (see p. 40) was critical to describing the genetic basis for inducible lactose utilization. Finally, these French scientists identified genes that acted in ways other than those conceived by Beadle and Tatum, who asserted "one gene–one enzyme." Monod, Jacob, and Lwoff had discovered regulatory genes as opposed to structural ones (i.e., genes that encode for enzymes or structural proteins). In addition, they proposed the existence of genes that act as targets for DNA-binding agents but do not encode proteins. Many of the current models for regulation of gene expression both in time and in "space" in all organisms and viruses have their basis in the operon model.

Temperature-sensitive mutations

Geneticists, from the beginning of their science, have recognized that genes do not act in a vacuum. The phenotype is the outcome of the interaction of the genotype and the environment. The term *genotype,* itself, conceals a host of interactions, since a gene functions only by virtue of its genetic milieu. Similarly, the term *environment* conceals a host of interactions; the meteorologists' "comfort index," for example, depends upon the interaction of temperature and relative humidity.

Temperature-sensitive mutations are a subcomponent of mutations known as conditional mutations. Temperature-sensitive lethals, in turn, are a subcomponent of temperature-sensitive mutations. Certain chlorophyll mutants in plants have been recognized since the 1920s as being temperature sensitive. In some instances, high temperature interferes with the normal synthesis of chlorophyll; in others, low temperature results in abnormal chlorophyll. Lethal mutations are frequently temperature sensitive; P. W. Whiting (1934) reported on such a gene in the parasitic wasp *Habrobracon juglandis* (at one time a popular genetic tool) (see Figure 7-7).

Temperature-sensitive lethals provide one of the most powerful tools in genetic research on the time and nature of gene action. Their power was often unappreciated by geneticists. For example, Th. Dobzhansky and Boris Spassky (1944), in an attempt to reveal the "manifestation of genetic variants in *Drosophila pseudoobscura* in

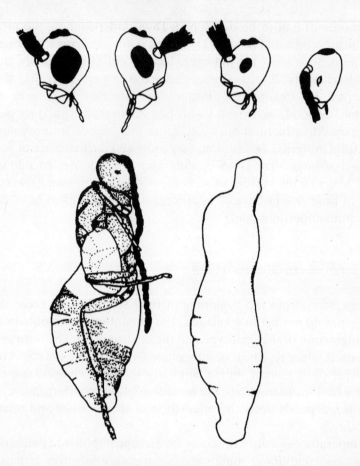

Kidney, k (eyes), and its allelomorphs small-eyes, k^s, and extreme-small, k^e.
A mutation to kidney, *k* (eyes), has been reported with a brief description of the mutant type. . . . When the wasps are bred at 30˚C compound eyes and ocelli are reduced in size or absent . . . and the majority of specimens are inviable, many dying in cocoons as elongate pupae often with small heads. . . . At lower temperatures kidney is highly viable and fertile in both sexes.

Figure 7-7. One of the earliest accounts of a temperature-sensitive lethal in "modern" genetic literature. The experimental organism is *Habrobracon juglandis*. (Top, left to right) A normal head and eye and three expressions of *kidney (k)* eyes. (Bottom) Two elongate, dead *kidney* mutants raised at 30°C. Beneath the drawings is P. W. Whiting's (1934, p. 269) original description of the mutant. (From Whiting, 1934, by permission of *Genetics*.)

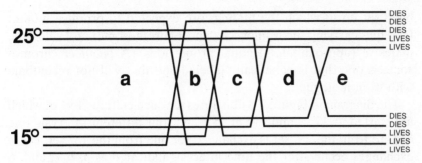

Figure 7-8. A diagrammatic account of the method for identifying the effective period during which a temperature-sensitive lethal mutation "acts." The diagram depicts five cultures that have been set up at 15°C (the permissive temperature) and 25°C (the restrictive, or lethal, temperature). Exchanges involving one culture of each series of five are made following periods *a, b, c,* and *d.* The diagram as drawn reveals that any culture that is exposed to 25°C during period *c* fails to yield flies carrying the temperature-sensitive lethal; conversely, cultures that pass period *c* at 15°C yield living, temperature-sensitive individuals. As does the upper half of Figure 7-5, the diagram shown here represents normalized developmental times; many more hours are required for development at 15°C than at 25°C.

different environments" tested the viability of flies homozygous for autosomes that were obtained from wild-caught flies. Cultures of flies were raised at $16^1/_2$°, 21°, and $25^1/_2$°C; in each culture, the expected proportion of wild homozygous flies was $33^1/_3$%. Dobzhansky and Spassky were successful in showing that, indeed, the proportions of wild-type flies in these cultures were different under different temperature regimens. Among the 49 chromosomes tested were two that were lethal to their homozygous carriers at $25^1/_2$°C, but essentially normal at 16°C (the proportions of wild-type homozygotes ranges from 25% to 28% rather than $33^1/_3$%). In this study, the demonstration of variation was the goal, and the goal was met by the experimental results. No thought was given to the further uses to which these (or any other) temperature-sensitive lethals could be put.

David Suzuki, a *Drosophila* geneticist, saw that temperature-sensitive lethals could be made to reveal the time during the individual's development at which the lethal played its fatal role. Figure 7-8 presents a schematic account of Suzuki's (1970) procedure. Imagine a sex-linked mutation in *D. melanogster* that is lethal to its hemizygous (male) and homozygous (female) carriers at 25°C but seemingly without effect at 15°C. Imagine, too, that cultures have been set up in which the matings are expected to produce mutant (say, flies carrying the Muller-5 balancer chromosome) and wild-type males in (approximately) equal proportions. (A *balancer* chromosome is one that is genetically labeled and that will not recombine with its homologue.)

The diagram in Figure 7-8 illustrates ten such cultures, five of which are started at 25°C, and five at 15°C. At four different intervals, one culture from the 25° set is exchanged for one from the 15° set. These exchanges occur after the time intervals indicated as *a, b, c,* and *d*. Whether wild-type males occur in these cultures is shown at the right. "Dies" means that no wild-type males survive; "lives" means that wild-type males are as numerous as Muller-5 males. These cultures reveal that those which were at 25°C during interval *c* yielded no wild-type males, whereas those that were at 15° during interval *c* did. Thus, the temperature-sensitive lethal exerted its effect during that interval. Further tests could narrow this interval to any desired width. The diagram is simplified in not revealing the greater time required for development at 15° than at 25°. The exchanges of cultures between temperatures would be governed by developmental stage (by larval instar, for example) rather than by time in hours (recall Figure 7-5).

To what further use in the study of gene action, other than localizing the effective time of action, have temperature-sensitive mutations been put? Norman Horowitz (1950) described their use in meeting a criticism leveled against the generally accepted one gene–one enzyme hypothesis of Beadle and Tatum. If, for example, indispensable functions (functions without which the organism will die) were never associated with unifunctional genes but, on the contrary, were always associated with genes that possess many functions, the experimental method by which *Neurospora* strains are screened for newly arisen mutations (see p. 20) would always recover mutations at loci whose genes have only a single function—hence, one gene–one enzyme.

The use of temperature-sensitive mutations allowed this criticism (or possibility) to be put to the test. The spectrum of nutritional requirements of temperature-sensitive and non-temperature-sensitive mutant strains of *Neurospora* are essentially identical. At the time when Horowitz discussed this problem, 405 of 484 known mutations (84%) had requirements that could be met by single chemical compounds; the other 16% had not been studied thoroughly. Further, available data suggested that about one-half of the temperature-sensitive mutants would not grow on either complete or minimal medium at the restrictive (high, or 35°C) temperature. That observation suggested that the average number of indispensable functions (those that are impossible to supplement) of each gene could not be high. Indeed, a formal analysis of the data—considering the experimental biases *against* demonstrating the one gene–one enzyme hypothesis—suggested that no reason existed for assuming that genes have more than one function. With the hard evidence that now exists (thus rendering hindsight more nearly perfect), one knows that some gene loci can be transcribed to yield different mRNAs (and, hence, different proteins) in different somatic tissues and organs. These are the few multifunctional gene loci allowed for by Horowitz's calculations.

Temperature-sensitive mutations have been of enormous utility in the study of essential functions. They provide the means by which the life-and-death differences between living and dying organisms can be identified and isolated for study. What is it that living organsims raised at the permissive (usually low) temperature lose when they are transferred to the restrictive (usually high) one? How quickly is it (that mysterious *it*) lost? Is it lost gradually, as if by dilution, or abruptly? In the hands of a competent, inquisitive investigator, these mutations provide access to a wealth of information. Two other important uses to which temperature-sensitive mutations have been put are the saturation of genetic maps with as many mutations as possible, and, in turn, the analysis of genetic organization that such saturation permits. Genetic mapping, one should note, is possible with temperature-sensitive mutations because the "matings" are performed at the permissive temperature while the progeny are tested under the restrictive temperature.

The study of the developmental biology of the bacteriophage T_4 provides examples illustrating these two uses of temperature-sensitive mutants. In addition to temperature-sensitive phage mutants (those that would infect bacteria and produce progeny phage particles at 25°C but were unable to do so at 42°C), the phage workers had a second class of conditional lethals (*amber* mutations) with which to work. *Amber* mutations are changes in the phage DNA that result in the mutating of a functional mRNA codon (say, UAC—the

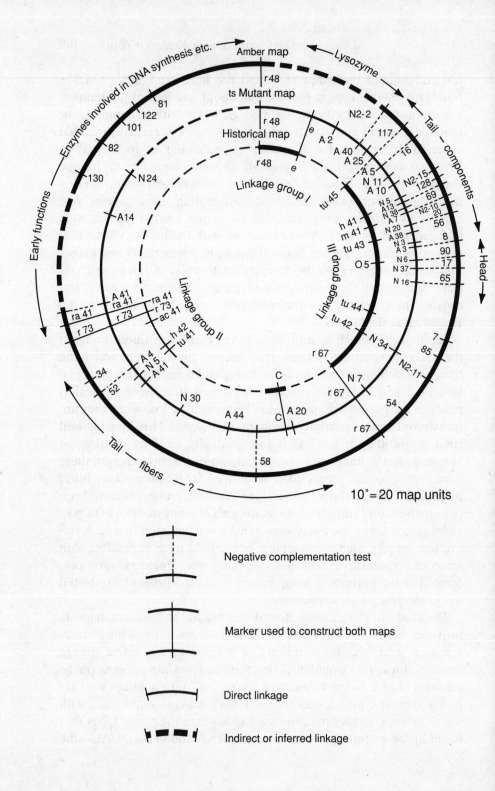

Amber map

ts Mutant map

Historical map

Lysozyme

Enzymes involved in DNA synthesis etc.

Tail — components

Early functions

Head

Tail — fibers

r 48

r 48

r 48

e

e

81

122

101

82

130

N 24

A 14

N2-2

117

A 2

A 40

A 25

A 5

N 11

A 10

N 5

A 13

N 38

A 7

N 20

A 38

N 3

A 3

N 6

N 37

N 16

16

N2-15

128

69

N2-10

20

56

8

90

17

65

Linkage group I

tu 45

h 41

m 41

tu 43

O 5

tu 44

tu 42

N 34

7

85

N2-11

54

A 41

ra 41

ra 41

r 73

r 73

ra 41

r 73

ac 41

h 42

tu 41

A 4

N 5

A 41

34

52

N 30

A 44

58

C

A 20

C

r 67

r 67

r 67

N 7

Linkage group II

Linkage group III

?

10° = 20 map units

Negative complementation test

Marker used to construct both maps

Direct linkage

Indirect or inferred linkage

codon for tyrosine) to a STOP codon (UAG, for example). That change may be sufficient to prevent the synthesis of progeny phage in an infected bacterium. However, a bacterial mutation may produce a tyrosine- (or other amino acid–) carrying tRNA that reads UAG incorrectly and inserts tyrosine into the growing polypeptide chain. These bacteria "suppress" the lethal aspect of the *amber* mutants; hence, *amber* mutations are conditional lethals: lethal in normal bacteria, non-lethal in bacteria carrying the suppressor mutation.

Neither temperature-sensitive nor *amber* mutations occur only in specific genes; rather, they affect any one of many sites within any gene or gene product. Thus, they allow the mapping of many sites within the phage's DNA chromosome. Figure 7-9 shows an early map that was constructed by Robert Edgar and R. H. Epstein (1965). From this map and its sequels, one can find the many correlations between "phenotypic" and more-molecular traits; for example, phage mutants that fail to infect the normal *E. coli* host but infect rare, newly arisen *E. coli* hosts (i.e., host-range mutants) map to positions occupied by genes involved in the synthesis of tail-fiber (see Figure 7-10) protein.

Through the use of *amber* and temperature-sensitive mutations, a virtually complete understanding of phage protein synthesis and their assembly into functional phage particles has been reached (Figure 7-10). This understanding required analyses of the amount of phage DNA in infected bacteria, information concerning the production or nonproduction of "viable" phage following the infection of bacteria with two or more nonfunctional particles, serological tests for various phage proteins, and electron micrographs of the inner contents of infected bacteria. Studies of this sort have also revealed which steps in the construction of a phage particle require enzymatic intervention and which ones proceed spontaneously. A mixture of purified phage heads from a bacterial culture that has been infected with a "tail" mutant and the purified bacteriophage tails from a culture infected with a "head" mutant will spontaneously generate phage

---

Figure 7-9. An early diagram of the T$_4$ bacteriophage circular genetic map showing the location of temperature-sensitive *(ts)*, amber *(am)*, and "classical" (Historical) mutants. When all mutants known at that time were accounted for in the diagram, they were found to be clustered with respect to both function and time of function. (Redrawn from a previously unpublished diagram obtained through the courtesy of Robert Edgar.)

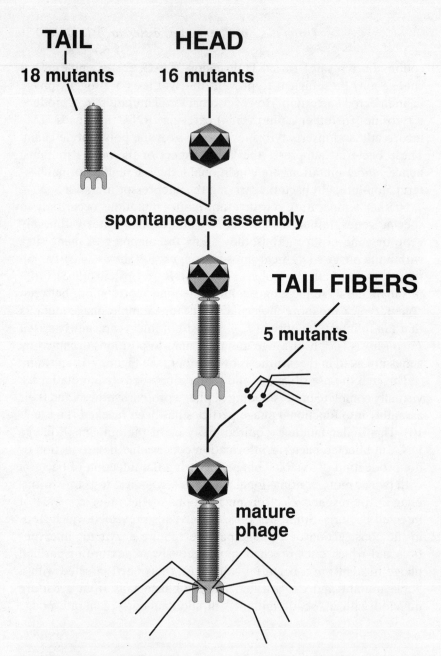

Figure 7-10. The assembly line that leads to adult T$_4$ bacteriophage. The heads, tails, and tail fibers are constructed separately and are then assembled. Also shown are the numbers of genes whose mutants are known to disrupt the indicated step in assembly.

particles capable of infecting bacteria and reproducing within them. The phage parts, in this case, assemble themselves into functional particles without enzymatic aid.

Heat-shock proteins

The exposure of cells to a heat shock induces, for a wide variety of organisms—from bacteria and yeast to flies and mammalian tissue-culture cells—special proteins known, for obvious reasons, as heat-shock proteins. A great deal of early information concerning these proteins was garnered by the study of the giant salivary cell chromosomes of *Drosophila melanogaster*. Consequently, a brief review of these chromosomes may be in order.

When the banded structures occupying the nuclei of salivary gland cells of a midge (*Chironomus* sp.) were first observed by E. G. Balbiani in 1881, they were not recognized as chromosomes. Their proper interpretation in *Drosophila* is commonly ascribed to Theophilus Painter of the University of Texas in 1933; it is more than likely, however, that H. J. Muller grasped their implication and urged Painter to study them (M. M. Green, personal communication). Merely because of their enormous size, giant chromosomes played an important role in understanding the physical rearrangements of chromosomes (e.g., inversions of chromosomal material that would surely escape notice in a metaphase preparation are clearly visible in salivary gland preparations) and in locating gene loci at specific sites on the chromosomes. During the 1950s, Wolfgang Beerman (also studying midges of the genus *Chironomus*) and Corowaldo Pavan (studying dipteran flies of the genus *Rhynchosciara*) discovered that, during larval development, certain regions of giant chromosomes grow in size (puff) at specific times only to "collapse" later (often with an obvious accumulation of additional DNA). Exposure of intact salivary glands to solutions containing tritium-labeled uridine revealed that these chromosomal puffs were sites of RNA synthesis (Figure 7-11).

The synthesis of heat-shock proteins by *Drosophila* salivary glands that have been briefly exposed to a temperature of 37°C can now be discussed, following an outline provided by Suzuki et al. (1981, pp. 747–748):

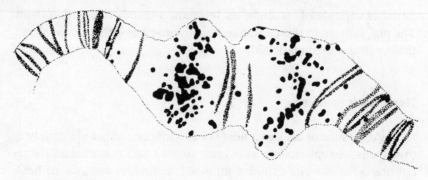

Figure 7-11. An autoradiograph revealing that tritium-labeled uridine accumulates preferentially at the site of puffs in giant salivary chromosomes (*Drosophila melanogaster,* in the present case). Because uridine is incorporated directly into RNA only, the dark grains in the overlying photographic film reveal that these chromosomal puffs are sites of intense RNA synthesis.

1. Nine regions on three major autosomal arms respond to the heat shock by forming puffs. These puffs become (microscopically) visible within a minute after the start of the heat shock.
2. Puffs that were present before the heat shock regress; they appear to be "turned off."
3. All RNA synthesis that preceded the heat shock is halted.
4. RNA synthesis that is specific for heat-shocked tissue (every tissue, not just the salivary gland) is induced. This RNA, whatever its source tissue, will hybridize in situ with the puff regions of salivary chromosomes.
5. Most protein synthesis ceases as a result of the heat shock; only those proteins known as heat-shock proteins are synthesized.

At one time, it appeared that this heat-shock syndrome provided an explanation for the connection between heat shocks and the induction of specific phenocopies. For example, the gene whose product is needed for completing a normal crossvein may act at a specific moment. If a heat shock is given to the flies at that moment (a shock that "normally" produces the *crossveinless* phenocopy), heat shock proteins are induced that, by binding with DNA, could turn off all normal gene functions, including that which is needed for the formation of the crossvein. According to this argument, a phenocopy results from the suppression ("turning off") of normal gene activity, not from some secondary or tertiary interaction among enzymes and various metabolites.

That simple hypothesis appears to be wrong. It is now known, for example, that heat-shock proteins are synthesized in response to many different stresses (heat, pH, anoxia, electromagnetic fields, and hydrogen peroxide, for example); furthermore, the entire repertoire of stress-induced proteins need not be evoked for any one stress. These proteins are ancient ones, a fact that is revealed by the similarity of their amino acid sequences regardless of their origin. Finally, in addition to possibly binding with DNA, these proteins may bind to and prevent the denaturation of other proteins—a role that has led some so-called heat-shock proteins to be renamed "chaperonins."

Although much about heat-shock proteins remains a mystery, some suggestions concerning their cellular functions have been made. For example, many nuclear proteins in mammalian cells are especially prone to damage by heat; these proteins tend to become insoluble and to form an abnormal insoluble protein aggregate. Heat-shock proteins bind with great and special affinity to such deformed protein molecules before aggregate formation. Thus, they can prevent the others from performing erroneous functions. Eventually, the "chaperoned" proteins can reassemble in their normal, preshocked conformations.

8 Fate Maps: Studying Development through the Use of Mosaics

In a paper delivered at the Sixth International Congress of Genetics in 1932, A. H. Sturtevant posed one question in two guises (p. 304): "One of the central problems of biology is that of differentiation— how does an egg develop into a complex many-celled organism? This is, of course, the traditional major problem of embryology; but it also appears in genetics in the form of the question 'How do genes produce their effects?'" Sturtevant went on to laud the technique of grafting that had been (and still is) used so effectively by experimental embryologists. Grafting, however, has certain limitations, some of which, he suggested, "can be avoided by the use of spontaneously occurring mosaics, such as gynandromorphs and somatic mutants." The use of mosaics should help in studying (1) in which parts of the body and (2) at what times of development specific genes are effective. "Where" and "when," as well as "how," are essential components of any study of gene action.

Sturtevant was Thomas Hunt Morgan's student and colleague who, as an undergraduate, realized that linkage data involving combinations of genes that overlap one another can be converted into a linear map containing all analyzed loci. For example, crossover frequencies may be known for the genes *a*, *b*, and *c*; *b*, *c*, and *d*; *b*, *d*, and *f*; and *a*, *f*, and *g*. If those frequencies are known, the data can be used to create a map that shows the relative positions of all seven of these genes: *a*, *b*, *c*, *d*, *e*, *f*, and *g*. The fundamental crossover data were obtained in each of the smaller studies by noting the frequency with which two alleles appeared among an individual's gametes in combinations *unlike* those with which they were associated at the individual's conception. If the parental gametes, for example were *AB* and *ab*, the amount of crossing over is calculated as the percent frequency of *Ab* and *aB* gametes among all gametes. For any pair of alleles, this percentage should not exceed 50%—the frequency expected under independent assortment. Expanding this type of reasoning from one dimension to two, Sturtevant was suggesting in 1932 that gynandromorphic mosaics might be used in mapping the developmental origins of the various portions of an adult fly.

Although the aim of fate mapping, as the procedure is now known, is to relate the distances between pairs of structures on the surface of a three-dimensional fly (*Drosophila melanogaster*, as a rule) to those between the responsible progenitor cells, sectors of mutant cells in bacterial colonies may provide insight concerning the physical disposition of the progeny of individual cells. Figure 8-1 illustrates that a mutant cell within a bacterial colony tends to produce a clearly defined sector of progeny cells, a sector that possesses remarkably sharp boundaries. If the mutation occurs at the first division of a one-celled "colony," the outcome may be a colony with two contrasting halves. However, the four cells that arise following the next division may shift so that the resulting colony exhibits alternating quarter sectors. In every case, whether the mutant cell arises early or late during the colony's growth, the outcome is a clearly defined sector of an appropriate size. This same rule applies to the small sphere of 32–64 nuclei that arise in a *Drosophila* egg before these nuclei migrate outward to form a mononuclear layer on the ellipsoidal surface of the developing egg. After a few additional divisions, cell membranes form and generate individual cells—one nucleus in each, but only after the nuclei have taken up their surface positions. As in the case of bacterial colonies, however, sharply demarcated mutant (and nonmutant) sectors are present among the cells that form at the egg's surface.

Sturtevant, several years before writing the paragraph cited at the start of this chapter, had discovered a mutation in *D. melanogaster* that (by the loss of an X chromosome) generated a high proportion of gynandromorphs, flies with sectors of both male and female tissues—often half and half. Using the same logic that led to the construction of linkage maps, Sturtevant reasoned that the closer two cells are to one another in the early development of a fly, the less

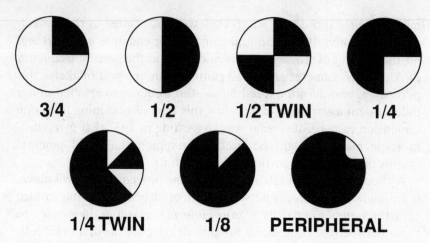

Figure 8-1. Colony patterns created by lactose-negative mutant cells (light areas) of *Escherichia coli* that arose after treatment with ultraviolet radiation. For present purposes, it is important to note that the mutant cells form well-defined sectors as the result of the geometry of colony growth. When the mutation occurs among the first few cells of a young colony, the resulting sectors reflect the accidental positioning of those early mutant and nonmutant cells.

likely the line separating male from female tissue is to fall between them. Labeling the two structures *A* and *B*, and using the letters *f* or *m* to indicate the sex of the sector within which each is found, Sturtevant simply calculated the frequency of *Af-Bm* plus *Am-Bf* flies among all flies examined. Small percentages indicate proximity of progenitor cells; large percentages indicate distant positions.

Using Sturtevant's yellowed notes, Antonio Garcia-Bellido and John R. Merriam found that his data did, indeed, yield a self-consistent map of body structures in the fly. In the meantime, a more efficient generator of gynandromorphs had been discovered: a ring-shaped X chromosome that is eliminated from one daughter cell, usually during the first nuclear division in virtually every fertilized egg. This elimination generally produces gynandromorphs (half male, half female) in which the plane separating the contrasting hemispheres is seemingly oriented at random relative to the main axis of the fly (Figure 8-2).

How does one convert the frequencies with which mosaic bound-

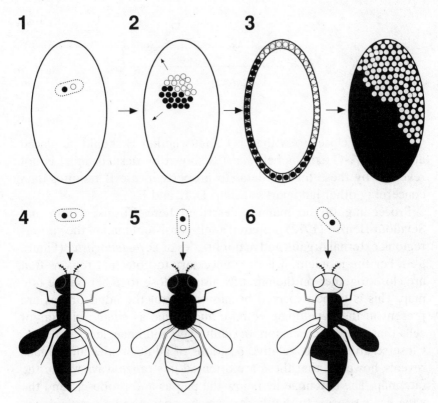

Figure 8-2. Patterns of male and female tissue in mosaic flies (gynandromorphs) following the loss of an X chromosome (open circles) in the first nuclear division following fertilization. At the top (1 and 2), the distribution of XX and XO nuclei are shown, followed by two diagrams (3) showing the distribution of XX and XO cells in the developing embryo. At the bottom (4, 5, and 6) are shown the distributions of male and female adult tissues as they are affected by the orientation of the first nuclear division. (Adapted with permission of Macmillan Magazines Limited and Seymour Benzer from Hotta and Benzer, 1972, *Nature* 240:527–535. Copyright 1972 Macmillan Magazines Limited.)

aries intercalate themselves between anatomical landmarks into a self-consistent map of those landmarks? Three landmarks—A, B, and C—can illustrate the (triangulation) procedure. Suppose that among 100 mosaic flies studied, A and B are found in dissimilar zones 12 times, A and C 16 times, and B and C 5 times. The results can be represented as follows:

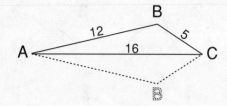

Clearly, B is closer to A than is C, but whether B should be placed above the A-C axis or below it (as shown by dashed lines) is not revealed by these limited data. To properly locate B requires data concerning other landmarks such as D, E, and F.

Proceeding in the manner described here, Yoshiki Hotta and Seymour Benzer (1972) prepared a self-consistent map of the various regions, external organs, and major bristles of *D. melanogaster* (Figure 8-3). For the moment, it is only necessary to note that regions that are close together on the adult fly also lie close together on the fate map. This is to be expected because many of the adult organs are present in the developing embryo and larvae as *anlage,* clusters of cells that are the precursors of adult eyes, antennae, legs, and such. Close scrutiny of the relative positions of adult bristles and organs reveals, however, that these are oriented in strange ways within the fate map: The antennae lie below the palpus and proboscis, and the wing lies posterior to the thorax, which appears to be upside down. The humerus, which on the adult fly lies near the upper portion of the anterior leg, lies in the fate map on the opposite side of the thorax. One must remember that insect development is not a simple matter. *Drosophila* wings, for example, develop inside out and are everted by fluid pressure much as the fingers of a rubber glove are everted by someone's blowing into the wrist.

The use of fate mapping in understanding where genes act took an enormous stride during the early 1970s. Benzer and his colleagues developed laboratory techniques for isolating behavioral mutants in *D. melanogaster*. They argued that the most effective means for obtaining mutations that affect behavior in flies is to mutagenize flies (using radiation or chemical mutagens) and select individuals that exhibit particularly obvious and inheritable changes in behavior. By subjecting flies to cleverly designed tests, these and subsequent workers isolated mutants among which are (chromosome and location are given in parentheses) *stuck* (4-0), *coitus interruptis* (1-22.1),

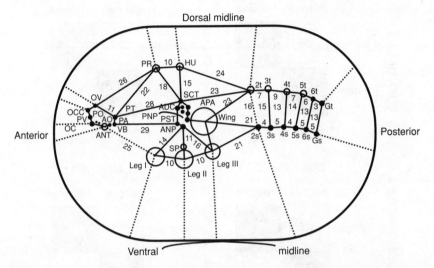

Figure 8-3. The embryonic fate map of adult structures in *Drosophila melanogaster.* The distances between nearby structures are measured in units called *sturts* (for A. H. Sturtevant, who first used this technique). One sturt equals 1% "recombination" between the structures and contrasting mosaic characteristics. ADC, anterior dorsal central bristle; ANP, anterior notopleural bristle; ANT, antenna; AO, anterior orbital bristle; APA, anterior postalar bristle; Gs, genital sternite; Gt, genital tergite; HU, humeral bristle; OC, ocellar bristle; OCC, occiput; OV, outer vertical bristle; PA, palp; PNP, posterior notopleural bristle; PO, posterior orbital bristle; PR, proboscis; PST, presutural bristle; PT, postorbital bristle; PV, postvertical bristle; SCT, scutellar bristles; SP, sternopleural bristles; VB, vibrissae; s, abdominal sternite; and t, abdominal tergite. (Adapted with permission of Macmillan Magazines Limited and Seymour Benzer from Hotta and Benzer, 1972, *Nature* 240:527–535. Copyright 1972 Macmillan Magazines Limited.)

sluggish (1-65), *hyperkinetic* (1-30.5), *flightless* (1-0.5), *easily shocked* (1-53.5), *tko* (1-1.0), *paralyzed* (1-52.1), *comatose* (1-40.1), *arrhythmic* (1-14), *short-period* (1-14), *long-period* (1-14), *phototactic* (1-36.6), and *drop-dead* (1-33-57). As the varied names imply (including *technical knockout, tko*), these mutations affect many aspects of the fly and its behavior: for example, response to light, ability to maintain a circadian (about 24-hour) rhythm, response to physical shock, and mating behavior.

The physical change that leads to an altered behavior is difficult to identify. The lesion may be in the brain, or elsewhere within the nervous system, or in a sensory organ, or in muscles that control the physical attitude of a fly. Fate mapping provides a powerful means

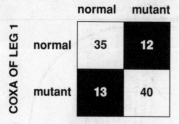

ANTENNA

	normal	mutant
COXA OF LEG 1 — normal	35	12
COXA OF LEG 1 — mutant	13	40

$\frac{12 + 13}{100} = 25$ PERCENT = 25 STURTS FROM ANTENNA TO COXA 1

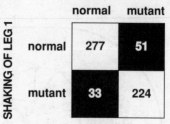

COXA OF LEG 1

	normal	mutant
SHAKING OF LEG 1 — normal	277	51
SHAKING OF LEG 1 — mutant	33	224

$\frac{84}{600} = 14$ STURTS FROM COXA 1 TO SHAKING FOCUS

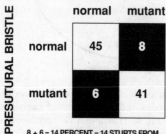

COXA OF LEG 1

	normal	mutant
PRESUTURAL BRISTLE — normal	45	8
PRESUTURAL BRISTLE — mutant	6	41

$\frac{8 + 6}{100} = 14$ PERCENT = 14 STURTS FROM COXA 1 TO PRESUTURAL BRISTLE

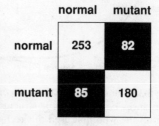

ANTENNA

	normal	mutant
SHAKING OF LEG 1 — normal	253	82
SHAKING OF LEG 1 — mutant	85	180

$\frac{167}{600} = 28$ STURTS FROM ANTENNA TO SHAKING FOCUS

PRESUTURAL BRISTLE

	normal	mutant
ANTENNA — normal	34	13
ANTENNA — mutant	16	37

$\frac{13 + 16}{100} = 29$ PERCENT = 29 STURTS FROM PRESUTURAL BRISTLE TO ANTENNA

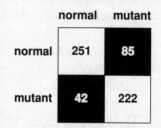

PRESUTURAL BRISTLE

	normal	mutant
SHAKING OF LEG 1 — normal	251	85
SHAKING OF LEG 1 — mutant	42	222

$\frac{127}{600} = 21$ STURTS FROM PRESUTURAL BRISTLE TO SHAKING FOCUS

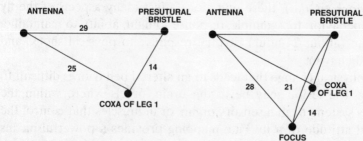

for locating the presumptive lesion in relation to physical landmarks. One treats normal and mutant behavior as one would normal and mutant morphology (a bristle, for example) and calculates the proportion of all instances in which one trait is normal and the other is mutant (see Figure 8-4). On the left of Figure 8-4 are the observations that lead by triangulation to the fate-map distances of three anatomical parts: antenna, presutural bristle, and coxa (proximal segment) of leg 1. The distances are given in *sturts* (1% "crossover"), so named in honor of A. H. Sturtevant. On the right are the fate-map distances (for the mutant *hyperkinetic*) of leg shaking (under etherization) for leg 1 (the anterior leg) and each of the three morphological landmarks. The map reveals that the responsible focus for leg shaking lies within the developing embryo, in a location that histological studies have revealed to be the origin of the ventral nervous system. Histological studies have revealed that the thoracic ganglia of the ventral nervous system are abnormal in *hyperkinetic* flies. It might be noted here that the focus of the leg-shaking lesion for each leg is independent of the other legs; each presumably implicates that leg's corresponding ganglion.

The next (and last) behavioral mutant to be discussed here is *dropdead*. These mutant flies seemingly develop normally, emerge from their pupae, but at an earlier-than-expected age they suddenly die. As Benzer (1973, p. 32) described it, "These flies develop, walk, fly, and otherwise behave normally for a day or two after eclosion. Suddenly, however, an individual fly becomes less active, walks in an uncoordinated manner, falls on its back and dies; the transition from apparently normal behavior to death takes only a few hours." The survival curve for these flies is given in Figure 8-5; the linear decline of the log of the proportion of survivors indicates an exponential decline resulting from random (cataclysimic, for the fly) events.

Figure 8-4. The use of triangulation to locate the causative lesion for a behavioral trait relative to the known locations of other traits. On the left are the observations that locate the relative positions of the embryonic structures that give rise to the antenna, presutural bristle (PST in Figure 8-3), and the coxa (proximal segment) of the first leg. On the right, the lesion responsible for *shaking* has been located with respect to these three landmarks; it lies beneath them, nearer the coxa than the other two markers. (Adapted from "Genetic Dissection of Behavior" by Seymour Benzer. Copyright © 1973 by Scientific American, Inc. All rights reserved.)

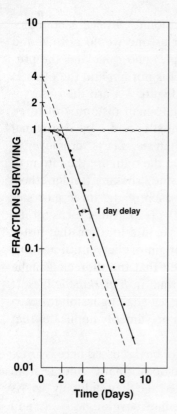

Figure 8-5. The survival curve of *drop-dead* mutant (•) and normal (○) males over a period of 10 days. Virtually all the normal males survive; *drop-dead* males begin dying on the second and third days and continue to do so until nearly all are dead by the 10th day. The straight-line decrease in the logarithm of the proportion of surviving flies suggests that these data might be interpreted as representing the inactivation of "targets"; this possibility is discussed in the text. (Adapted from Hotta and Benzer, 1972.)

The analysis needed in locating the causative lesion for *drop-dead* proved to be more complex than for *hyperkinetic*, but solvable nonetheless. The lesion was located within the fly's brain. Actually, two foci were involved, either one of which alone is capable of sustaining the fly in apparently good health. When both foci failed, however, the fly died. Microscopic examination revealed the degeneration of the brain that accompanied the fly's dying; flies that did not die had normal-appearing brain tissue. These histological observations confirmed the conclusions that had been reached by analyzing fate-map data.

The data presented in Figure 8-5 invite still further comment. Survival curves, with or without an initial shoulder, that become linear when plotted on semilog graph paper may reveal the number of "targets" that must be destroyed in order to kill an individual. The simplest such curve is, of course, one that declines linearly from the start (100% survival)—that is, the log of the proportion of survivors declines linearly with time or with exposure to a noxious agent such as X-radiation, ultraviolet light, heat, or a toxic chemical. A survival curve of this sort suggests that death follows the destruction of a single target. If the target is not destroyed by the first period of exposure, the probability of hitting it during the second period remains unchanged. In such studies, 37% survival has special meaning because it represents the proportion of individuals whose "targets" have not been "hit" when the average number of hits per target is one (the proportion suffering zero hits $= e^{-m}$ where m is the number of hits per target; if $m = 1$, $e^{-1} = 0.37$, or 37%).

What if there are multiple targets, all of which enjoy the same probability of being hit? Following a scheme given by Suzuki and co-workers (1981, p. 246), one can see that

1. the probability that the first target is hit $= 1 - e^{-d}$ (where d equals *dose*, or the average number of hits per target);
2. the probability that the second target is hit $= 1 - e^{-d}$;
3. the probability that the third target is hit $= 1 - e^{-d}$; and, finally,
4. the probability that the kth target is hit $= 1 - e^{-d}$.

The probability that all k targets have been hit equals $(1 - e^{-d})^k$, or approximately (i.e., ignoring higher terms) $1 - ke^{-d}$. The proportion of individuals possessing at least one surviving target (and, hence, surviving) $= 1 - (1 - ke^{-d}) = S = ke^{-d}$. Upon solving we see that

$$\ln S = \left(\ln k\right) - d.$$

When $d = 0$, the intercept of this equation equals $\ln k$—that is, the intercept reveals the number of targets (if, indeed, target theory is an appropriate theory).

Returning to Figure 8-5, we see that the survival curve is consistent with the need to destroy *four* targets in order to kill a *drop-dead* mutant. However, the fate-map analysis revealed only two foci. It appears possible, then, that the space between the two curves (that which fits the observed data and the parallel one which is expected to correspond to two targets) reveals the length of time a doomed fly (both targets or foci destroyed) moves about before collapsing and dying. The line suggesting the existence of four targets, in this case, does so by coincidence. On the other hand, if no histological evidence for brain damage can be found in surviving flies (some of which are already doomed), then death may follow rapidly upon the destruction of the last of four targets. Each focus, in this case, would actually be an undetected doublet of some sort. Whether it is, is unknown.

Somatic mosaics and compartments

Within a year of Muller's (1927) announcement that the frequencies of gene mutation are increased by exposure of *Drosophila* to X-rays, J. T. Patterson, of the University of Texas, was employing this new tool in an attempt (1) to induce mutations in somatic cells and (2) to use these mutations in an effort to trace cell lineages during embryonic and larval development. Patterson's numerous experiments involved several gene loci at which mutations have visible phenotypic effects in the adult fly. The *white* locus, which is on the X chromosome, has been chosen for illustrative purposes (Figure 8-6). One might, of course, irradiate wild-type male embryos and larvae, anticipating that the w^+ locus in some cell might mutate to w, *white* eye. One might also—and with greater success—irradiate w^+/w heterozygous female embryos and larvae, with the expectation that a small portion of the w^+-bearing X chromosome will be lost, thus revealing a patch of *white* (male tissue) facets in the compound eye. The latter procedure, as one might anticipate because a chromosome may be broken in many places, generates more white patches than does the one that relies on specific site mutations. (*Mitotic crossing over* [p. 171] may also produce white patches.) Figure 8-6 illustrates four eyes, each of which contains an area consisting of white, rather than red, facets.

The size and frequency of mutant patches of adult tissue depend upon the time during its development when an embryo is irradiated, just as sector size in a bacterial colony is influenced by the time when a gene mutates. Early embryos have few cells (and early *Drosophila* larvae have few cells in the *anlagen* that eventually give rise to adult organs and tissues); consequently, they are not likely to be "hit," but, when they are, they give rise to a large patch of mutant tissue because they have many descendants. Later in development, many more cells are present, so one or more are quite likely to be hit, but each gives

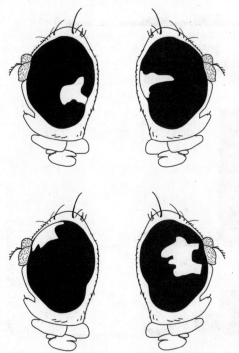

Figure 8-6. White patches in the otherwise red eyes of offspring of w^+/w heterozygous females that were exposed to X-radiation during early larval life. Most such mutant patches result from the loss a small portion of the w^+-bearing chromosome or from mitotic recombination. (After Patterson, 1929.)

rise to a rather small patch of mutant tissue. (A similar pattern exists in the case of irradiated mouse embryos; Russell and Russell, 1954: When mouse embryos are irradiated very early, most are not affected; those that are usually die—early in embryonic development. When embryos are irradiated at a later stage, most have cells here or there that are hit [mutated]; consequently, many mice that are irradiated late are born with crippling, but not lethal, defects. Presumably, the same is true for human embryos that have been exposed to X-rays or other mutagens.)

The dependence of the size of mutant areas on flies that were irradiated as embryos and larvae is shown in Figure 8-7. Seven exposure periods are shown (0–12 hours after egg laying, 12–24 hours, 24–36 hours, 36–48 hours, 48–60 hours, 60–72 hours, and 72–84 hours). The mode or median number of affected (mutant) facets of the compound eye for successive exposure intervals moves steadily to the left, culminating eventually in numerous instances in which a single facet is expressed as the mutant phenotype.

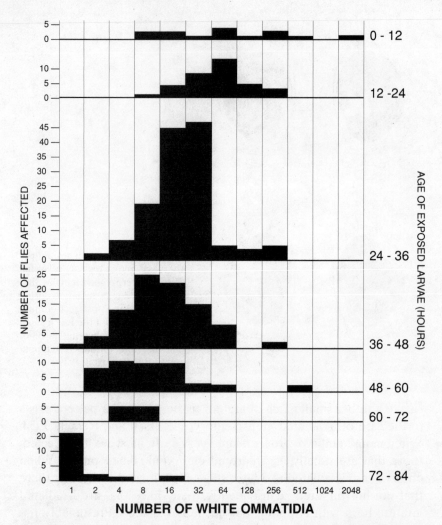

NUMBER OF WHITE OMMATIDIA

Figure 8-7. The relation between the size (number of affected ommatidia) of the mutant sectors (see Figure 8-6) and the age of the exposed *Drosophila* larvae. The number of cells in the eye *anlage* of young larvae is small; therefore, mutant cells, although rare, give rise to large sectors. The number of cells in the eye *anlage* of older larvae is large; therefore, although mutant sectors are numerous, they tend to be small. (Compare with the mutant sectors of bacterial colonies illustrated in Figure 8-1.) Because different numbers of flies were examined in the many experiments on which these seven graphs are based, the relative numbers do not represent relative frequencies. Note that the horizontal scale progresses by successive doublings, not arithmetically. (After Patterson, 1929.)

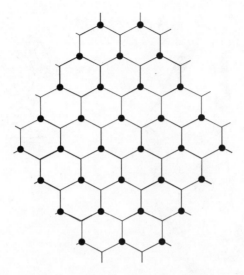

Figure 8-8. A diagrammatic representation of the arrangement of ommatidia and hairs on the surface of the compound eye of *Drosophila melanogaster*. Each ommatidium terminates on the surface in a hexagonal lenslike facet. The arrangement of these facets is exceedingly accurate. Each facet abuts on three small hairs (filled circles). Because of the geometry of hexagons, each row of hairs is displaced relative to the two adjacent rows, which, in turn, are matching rows. Many of the mutants known for *D. melanogaster* affect the physical structures (as well as the pigments) of the fly's compound eye.

In the absence of contradictory data, most geneticists tend to regard development as an unfolding of cellular potentialities that resembles in many ways the "unfolding" of patches of white facets in the compound eye of an irradiated fly: an event occurs (mutation of the w^+ locus, or the loss of a small segment of the w^+-bearing chromosome in a w^+/w heterozygous female) that determines that all descendant cells that come to lie within the otherwise red-pigmented compound eye will be white (lack pigment). This view, although both temptingly simple and frequently true, is not adequate to explain many developmental details. Intercellular chemical communication leads many diverse cell lineages to a common end.

The compound eyes of insects are marvels of intricate construction. Those of *Drosophila* are no exception. Figure 8-8 shows a portion of the surface of the compound eye of *D. melanogaster*. The surface consists of nearly 1000 hexagonal facets arranged with virtual crystalline accuracy. At three of the six vertices of each hexagon, there

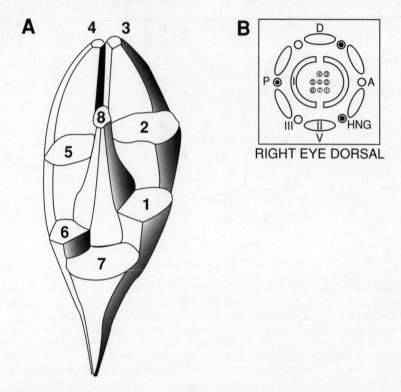

Figure 8-9. (A) A cutaway view of the eight photoreceptor cells in the ommatidium of *Drosophila melanogaster*. (After Ready et al., 1976, by permission of Academic Press and Seymour Benzer.) (B) A cross-section of an ommatidium, showing the eight photoreceptors (numbered) surrounded by primary (I), secondary (II), and tertiary (III; open circles) pigment cells and by nerve cells (HNG) leading to the small hairs (◉). A, P, D, and V are abbreviations for anterior, posterior, dorsal, and ventral. (After Lawrence and Green, 1979, by permission of Academic Press and Peter Lawrence.)

is a small hair (represented as solid circles in the figure). Students in beginning genetics classes are familiar with these eyes, whose color mutants (*white, eosin, vermillion, sepia,* and many others) provide materials for confirming Mendelian ratios in mono- and dihybrid crosses. These students are also familiar with the "dark spot" on the wild-type eyes, the spot that seemingly moves about within the eye— always pointing upward into the microscope. This spot, of course, is "formed" by those ommatidia down which the student is peering at any moment. Eye texture, as well as eye color, is altered by many

mutants (*rough, echinus, eyeless,* and others); these mutant phenotypes characteristically possess ill-shaped facets and misplaced or
missing hairs.

Within each of the ommatidia is a cluster of eight photosensitive
cells (Figure 8-9A). Both their proximity and their similar functions
suggest that they may represent a clone of cells descended, through
three cell divisions, from a single progenitor cell. This, however, has
been shown *not* to be the case.

Figure 8-9B represents cells of an ommatidium other than the eight
photoreceptors. The two semicircular cells surrounding the eight photoreceptors are the primary pigment cells; they serve to prevent light
that enters one ommatidium from activating photoreceptors in other
ommatidia. (White-eyed mutant *Drosophila* lack this pigment and, as
a result, have poor visual acuity—a defect that results in the effective
sterility of white-eyed *D. subobscura* males.) The six large ovals, each
one shared with a neighboring ommatidium, are secondary pigment
cells. The smaller circles represent tertiary pigment cells (open
circles) and nerve cells (solid dots within circles).

Peter Lawrence and Sheila Green (1979; see also, Ready et al.,
1976), by taking advantage of recombination within somatic cells
(mitotic recombination) and by using chromosomes both of which
contained mutations (but different ones) within the *white* locus, were
able to generate cells containing red pigment within an otherwise
white eye. The recombination events can be diagrammed as follows:

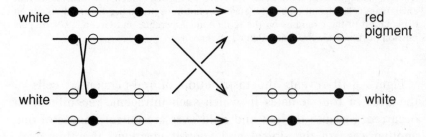

The distribution of red pigment within the ommatidia of otherwise
white-eyed females (i.e., XX flies) reveals, then, the fate of the
descendants of the cell in which this rare, intralocus recombination
occurred. (Note that the production of red pigment requires that one
of the two chromosomes carry two wild-type [filled circles] "sub-
genes" at the left.)

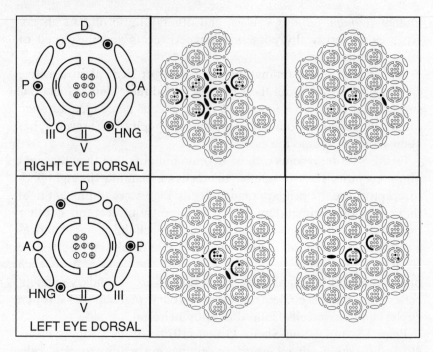

Figure 8-10. Observations demonstrating that neither the eight photoreceptors nor the various pigment cells of an ommatidium are monoclonal—i.e., related by immediate descent from a single precursor cell. By rare intragenic recombination, it is possible to obtain otherwise-white-eyed flies in which a single cell has acquired the ability to synthesize eye pigment (see text). The descendants of that single cell, all normally pigmented but lying within an otherwise unpigmented (white) eye are typically distributed among several ommatidia without revealing any tendency to form clones that encompass cells of one type only (such as photoreceptors). See Figure 8-9B for an explanation of the diagrams on the left. (After Lawrence and Green, 1979, by permission of Academic Press and Peter Lawrence.)

Figure 8-10 reveals the distribution of red-pigmented cells in the eyes of four females in which such intragenic recombination occurred. Because the left and right eyes are mirror images of one another (as are the dorsal and ventral positions of either eye), sketches resembling Figure 8-9B are shown at the left of Figure 8-10 to aid in orientation. As the patterns of pigment distribution reveal, in no instance are all eight photoreceptors of an ommatidium pigmented. This is true even of the eye (middle, top of diagram) in which the most pigmented cells occur; the pigmented cells are scattered seemingly at random among seven ommatidia. Cells of a crossover clone, all descended of necessity from a single "parental" cell, become photoreceptors, primary pigment cells, secondary pigment cells, tertiary pigment cells, and nerve cells independent of any genetic program that their lineage may have possessed at the outset.

Where, then, does the eventual fate of these eye cells reside? Apparently, in cell-to-cell communication. Using the scanning electron microscope, one can see a transverse groove in the "embryonic" tissue (*anlage*) that is to become the adult eye. This groove (*morphogenic furrow*) moves across the *anlage* and, as it does so, the disorganized cells lying in front of it become organized as the groove moves on so that, behind the advancing morphogenic furrow, the cells form a virtually crystalline lattice (Figure 8-11). The means by which this visible organization occurs is not understood. Suffice it to say that antibodies that have been generated against *Drosophila* neural proteins can, under proper experimental conditions, be seen attaching first to cells in the morphogenetic furrow and then in the sequence 8, 2 and 5, 3 and 4, 1 and 6, and, finally 7 (see Figure 8-9A); this sequence reveals the sequential synthesis of the target proteins. By the time the antibody binds to photoreceptors 1 and 6, however, it no longer binds to 2, 5, or 8. Furthermore, when it binds with photoreceptor 7, it no longer binds to any other of the eight photoreceptor cells. Thus, the target protein is synthesized—and then lost—in a precise pattern. These observations merely hint at the molecular research that will be needed to understand the development of any one of the 800 ommatidia that every fly constructs for one compound eye in less than a week's time.

Clearly, eight photoreceptors of each ommatidium do not possess a cell lineage that traces back (three generations) to a single progenitor cell. On the contrary, they often trace back to two, three, or maybe four cells. Comparable noncorrespondence between cell lineages and fates had earlier (in studies of wing structures) led developmental geneticists to the "compartment hypothesis" (Crick and Lawrence, 1975; Lawrence and Morata, 1976). An attempt has been made to represent cell lineages and compartments diagrammatically in Figure 8-12.

In Figure 8-12a, two cell lineages of two cell generations each are shown to contribute cells to two compartments, A and B. The compartments, rather than lineage, determine the anatomical or morphological characteristic of the cells, much as we have seen in the case

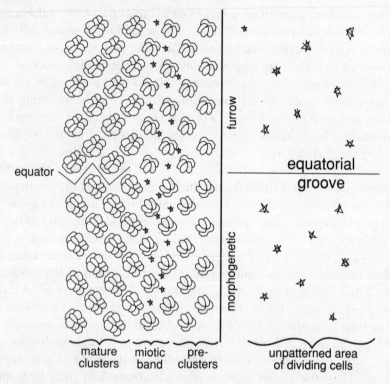

equator

furrow

equatorial
groove

morphogenetic

mature   miotic   pre-
clusters   band   clusters

unpatterned area
of dividing cells

Figure 8-11. A diagram illustrating the transformation of irregularly arranged cells lying in front (right) of the morphogenetic furrow into regularly arranged and positioned clusters of cells behind this moving furrow (left). Because clones are not involved in the formation of these clusters, the responsible recruiting signal must pass from cell to cell without regard to cell lineage. (After Ready et al., 1976, by permission of Academic Press and Seymour Benzer.)

with the photoreceptors and primary, secondary, and tertiary pigment cells and nerve cells in the *Drosophila* ommatidium. The information provided in Figure 8-12a is inadequate to reveal either the cell lineages or the compartments; effectively, the figure contains no information.

In Figure 8-12b, a mutation has occurred in the "stem cell" of lineage 2, resulting in a visible change of cell-2 progeny in the organism's body. This change can be detected in both compartment A and compartment B.

Figure 8-12c illustrates the consequences of a mutation in one

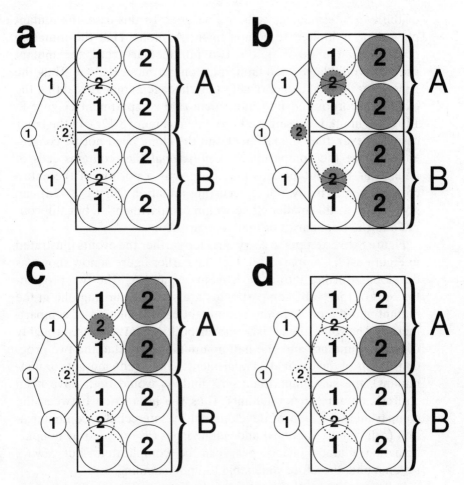

Figure 8-12. Diagrammatic representation of cell lineages and compartmentalization (compartments A and B). Two precursor cells (1 and 2) are shown undergoing two cell divisions, thus producing four descendant cells each. These eight cells are recruited by cell-to-cell signals into forming two compartments, A and B. Diagrams b, c, and d reveal the distributions of visibly mutant cells within and between the compartments when the mutation occurs at different times in lineage 2. Early mutations involve cells in both compartments (compare with Figure 8-10); later ones are limited to a single compartment but never involve the entire compartment because of cells of lineage 1.

of the two intermediary cells of lineage 2. In this case, the mutant phenotype is limited to compartment A, alone. Had the mutation occurred in the other of the two intermediary cells, the mutant progeny would have been limited to compartment B alone. Thus, the genetic marking of clones of cells (e.g., lineage 2) may reveal that the clone is incorporated into more than one compartment (Figure 8-12b) or is limited to a single compartment (Figure 8-12c). In the latter case, however, the compartment and the clone need not be congruent; because the compartment is *polyclonal* (i.e., contains cells of lineage 1), the mutant cells may form only a portion of the entire compartment. Finally, a late-occurring mutation (Figure 8-12d) can give rise to a still smaller clone within compartment A, but still confined to the boundaries of that compartment.

Figure 8-13 attempts to carry one step further the events illustrated in Figure 8-12. Compartment A of the earlier figure is now shown as developing further into compartments A_1 and A_2. The two progenitor cells of the "sub" compartments are shown as arising one in the mutant and the other in the nonmutant cell population of compartment A. The consequence is as shown: both A_1 and A_2 consist roughly of one-half mutant and one-half nonmutant cells. Had the two progenitor cells of the original compartment A both been mutant (or nonmutant), the subcompartments A_1 and A_2 would both have been entirely mutant (or nonmutant). Thus, the interaction between the compartmentalization of development by intercellular communication (cell-cell interactions) and cell lineages (as revealed by nonparticipating mutant phenotypes) can be complicated—but always rational and, therefore, amenable to understanding.

Following an outline provided by Lawrence and Morata, we can enumerate a few features of the compartment hypothesis. Compartments, although not uniclonal, are derived by descent from a few (polyclonal) neighboring cells. As shown in Figure 8-13, a compartment may become divided into subcompartments. Indeed, this process of subdivision probably continues throughout development as finer and finer "decisions" are made.

The fate of a polyclone that then gives rise to a compartment may be (i.e., we are surmising) determined by a "selector" gene's reacting to an intercellular cue. Once "turned on" the selector gene remains active within the cells' lineages, thus leading all affected lineages to the compartment's commonality. Cell surface proteins are among the

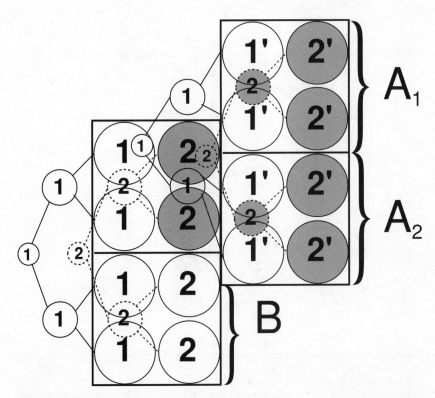

Figure 8-13. An extension of Figure 8-12, showing that compartment A undergoes further developmental change, giving rise to recognizably different subcompartments, A_1 and A_2. Two cells of the earlier compartment A undergo changes that lead to the subsequent subcompartments. If the two consist of one mutant and one nonmutant cell, as shown, the mutant phenotype will be exhibited in both subcompartments. If, however, the two cells of compartment A were both derived from a mutant sector, both subcompartments would be entirely mutant in appearance. Conversely, if both A cells were nonmutant, subcompartments A_1 and A_2 would be wildtype in appearance.

characteristics affected by selector genes. Cells of two compartments (say, eye and antenna in the case of *D. melanogaster anlagen*) when disassociated and then mixed in a culture dish will reassociate in two groups that reflect the original compartments. Within a compartment, information concerning size and position appears to be available to individual cells. Thus, a wing compartment that forms a portion of the entire wing in *Drosophila* maintains its same size relative to the rest of the wing whether the individual constituent cells are large or small;

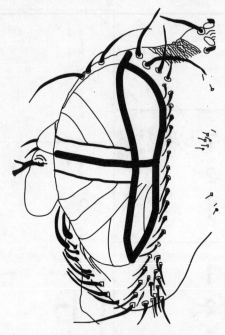

Figure 8-14. Sectors of the compound eye of *Drosophila melanogaster* as revealed by early induced mitotic recombination and chromosome loss (fine lines, see Figure 6-10) and developmental compartments. The larger compartments outlined by heavy lines were identified using recombinant cells marked by rapid (normal) growth in an otherwise mutant background; these compartments have some, but not all, boundaries that are coincident with those observed following early recombination. (After Baker, 1978.)

cell number compensates for cell size—cell number being, of course, a function of the number of cell divisions within the given polyclone.

The compartment hypothesis (as can be inferred from Figures 8-12 and 8-13) makes the interpretation of mutant mosaics difficult but not necessarily impossible. In Figure 8-14, the numerous fine lines delineate patches of mutant ommatidia observed by Hans Becker (1957) when somatic recombination was induced by exposing newly hatched larvae to X-radiation. The sectors were consistent enough from one affected individual to the next to enable him to conclude that the bottom half of the fly's eye arises from an early collection of eight cells (corresponding to the numbered sectors in Figure 6-10A). Later work has shown that the upper and lower halves of the compound eye form two compartments, a demonstration that does not negate the finer-grained analysis by Becker. The line separating the upper and lower portions of the compound eye is nearly—but not perfectly—coincident with a line above and below which the arrangement of the eight photoreceptors of individual ommatidia are mirror images of one another (see Figures 8-10 and 8-11). Still another compartmental boundary divides the compound eye into a large anterior and a smaller posterior portion. This boundary is not associated with any obvious landmark change in the compound eye.

A puzzling complication

That the present text is primarily designed for the nonprofessional modern (= molecular) geneticist (whether that person is or intends to become a biologist, engineer, economist, or agriculturist) does not exclude the possibility that the reader enjoys puzzles. The one to be described concerns four bristles (left anterior, right anterior, left posterior, and right posterior) that are located on the scutellum (small shield), a triangular portion of the thorax; the scutellum is the posterior part of the thorax (see Figure 6-9). Embryologically, the scutellum develops as two halves, one on either side of the embryo, which then migrate to the dorsal midline and fuse. Each half of the scutellum appears to begin as a single cell; the two halves are no more related to one another than any two cells of the early embryo.

If chromosomal loss (or mitotic recombination; see p. 171) is induced by X-radiation in young female larvae that are heterozygous for the mutant gene *forked,* adult flies are obtained some of whose bristles are twisted and misshapen—the characteristic of the forked phenotype. If the dose of radiation is low and (as a consequence) the proportion of flies with mutant patches is small (1 in 20, for example), the misshapen bristles occur either on the left or on the right side of the fly—not on both sides. This pattern is consistent with the supposition that all mutant cells are descended from a single cell in which the genotypic change was induced by radiation. Exposure of early larvae generally results in both the anterior and posterior scutellar bristles' being forked; left and right sides are affected with equal probability. Exposure of late larvae generally results in one (not two) forked scutellar bristles. This one bristle may be either anterior or posterior with equal probability; again, left and right sides are affected equally frequently. These results (which were reviewed earlier on p. 121 and Figure 6-9) suggest that, as the half-scutellum (right or left) develops from the initial cell, the cells that are destined to form the anterior and posterior bristles have a common progenitor. A chromosomal alteration that is induced in this progenitor cell results in both an anterior and a posterior forked bristle. Later, when the cell lineage splits and the anterior and posterior bristles have separate lines, an alteration affects only one or the other, not both. The data suggest that the two, now-separate lineages leading to the anterior and posterior bristles are equally susceptible to X-ray–induced chromosome loss.

With this brief review of the radiation biology of scutellar bristles, we can now turn to a mutant, *scute*. Numerous mutant alleles at the *scute* locus are known; their study forms a fascinating story in itself. Here, however, we are interested in the *scute* phenotype: *scute* mutations prevent the development of certain (not necessarily all) of the fly's large bristles; in the vernacular, they "remove" bristles.

Restricting our attention to the scutellum alone, we can concentrate on those flies that have only two bristles (rather than four)—one anterior bristle and one posterior one. The X-ray data cited above would suggest that an early event that "removes" a bristle would occur either in the left or the right half of the scutellum; hence, the two missing bristles would be missing from the same side of the fly. Consequently, the two bristles (anterior and posterior) that are present would lie on the same side. These expectations are not borne out by the observations!

Alternatively, one might calculate the probability that an anterior bristle is present (P_a) or absent $(1 - P_a)$. Similarly, one can calculate the probability that a posterior bristle will be present (P_b) or absent $(1 - P_b)$. The probability that both bristles occur on one side equals $P_a \cdot P_b \cdot (1 - P_a) \cdot (1 - P_b)$. The probability that the anterior bristle will occur on one side (say, the right) and the posterior one on the other (the left) equals $P_a \cdot (1 - P_b) \cdot P_b \cdot (1 - P_a)$. Clearly, the two products are identical; "same side" and "diagonal" patterns should be equally frequent. This expectation is also not borne out by the observations!

The patterns of present and missing bristles commonly seen on the scutella of *scute* flies are similar to those enumerated in Figure 8-15. Among flies that possess one anterior scutellar bristle and one pos-

R\L	+/+	+/−	−/+	−/−	Total
+/+	301	51	107	8	467
+/−	37	34	49	16	136
−/+	98	48	224	63	433
−/−	5	18	42	70	135
Total	441	151	422	157	1171

OBSERVED

R\L	+/+	+/−	−/+	−/−	Total
+/+	171	57	165	55	448
+/−	57	19	55	18	149
−/+	165	55	158	53	431
−/−	55	18	53	18	144
Total	448	149	431	144	1172

EXPECTED

Figure 8-15. The distributions of left (L) and right (R) anterior and posterior scutellar bristles as observed in certain *scute* flies and as expected if presence (+) and absence (−) occurred at random. The difference in totals is the result of rounding errors. The puzzling aspects of these data, as explained in the text, are (1) the seeming independence of presence or absence of anterior bristles with respect to that of posterior bristles (marginal totals) and (2) the low number of flies possessing both bristles on one side and none on the other. The early induction of somatic recombinants by X-radiation reveals (Figure 6-9) that the two bristles (anterior and posterior) on one side share a common lineage, and, therefore, their presence or absence should be even more highly correlated than otherwise expected.

terior one, the diagonal pattern in which the anterior bristle is on one side and the posterior one on the other is much more common than the "same-side" patterns in which the two bristles are on the same side.

The probability that an anterior bristle is present (P_a) need not be the same as the probability that a posterior bristle will be present (P_b). In Figure 8-15, $P_a = 50.9\%$ (nearly half and half), and $P_b = 75.0\%$. The marginal totals suggest that anterior and posterior bristles are present or absent at random with respect to one another. Furthermore, the individual cells of the figure (which represent the 16 patterns with which the four scutellar bristles can occur) reveal that the flies with diagonal patterns of one anterior and one posterior bristle occur in numbers corresponding to the expected ones. The asymmetrical flies exhibiting the "same-side" pattern are not seen in their expected numbers. Their numbers are much smaller than expected.

Here is a convenient place to cease discussing the details of this puzzling complication. Those who wish might refer to Wallace, 1983;

before doing so, however, it may be amusing to see what sorts of arithmetic manipulations will generate the seeming correlation (a *descriptive,* not an *explanatory,* term) between left and right sides of the fly while leaving the marginal totals unchanged. And one must not forget the physical isolation of the progenitor cells of the two halves of the scutellum; they lie on opposite sides of the developing embryo.

9 Transposable Elements

In a child's version of his autobiography (1969, p. 7), Erich Kästner wrote: "Wer von sich selber zu erzählen beginnt, beginnt meist mit ganz anderen Leuten. Mit Menschen, die er nie getroffen hat und niemals treffen wird. Mit Menschen, die schon lange tot sind, und von denen er fast gar nichts weiss." Loosely translated, Kästner said that he who wishes to tell about himself (as Kästner had set out to do) had best begin with entirely different persons—with persons he never met and never will meet, with persons long dead whom he never will know. This admonition applies to the subsequent discussion of transposable elements. To understand transposable elements as they exist in maize or *Drosophila,* for example, one must start with DNA, the physical basis of heredity. Why is DNA *the* genetic material? As the bearer of genetic information, what physical attributes proved to be useful and therefore became general properties of DNA? What ancillary problems were posed by these attributes? An understanding of transposable elements must begin with an attempt to understand problems that existed at the dawn of life. These problems are those associated with structure, function, and the transmission of information (i.e., accurate replication).

In the beginning . . . and later

With the improved vision that accompanies hindsight, we can associate molecular structures and functions more clearly today than was

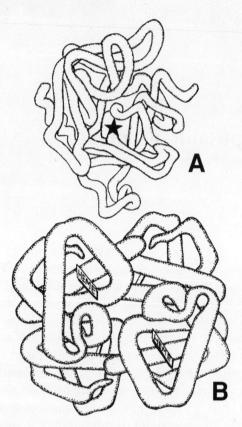

Figure 9-1. Two complex protein molecules—chymotrypsin (A) and hemoglobin (B)—whose ability to function depends on the creation and maintenance of an active site. The amino acids near the starred site at the center of the chymotrypsin molecule allow this enzyme to cleave polypeptide chains; the names and locations of these amino acids are histidine (57), aspartic acid (102), and serine (195). In the case of hemoglobin, otherwise distant amino acids are brought into proximity by the folding of each polypeptide chain; these retain the heme group, which in turn holds or releases oxygen.

possible even a half-century ago. Proteins can serve as an illustration. They perform their special functions as catalysts (enzymes); as structural molecules providing physical support to cells; as carriers transporting small molecules by virtue of their large size and complex physical structures; or as regulators of gene expression. Two complex protein molecules are shown in Figure 9-1. At the top (A) is a molecule of chymotrypsin, a protease. The cleavage of the polypeptide chain occurs at the enzyme's *active site,* the starred region near the center of the drawing. The rest of the large molecule probably provides the shape, physical strength, and hydrophilic properties that are needed for carrying out its role in digestion. The lower diagram is a molecule of hemoglobin, a complex of four polypeptide chains, two alphas and two betas. Each polypeptide chain is folded so that an oxygen-transporting heme group is held in place by several nearby amino acids. The folding of the polypeptide chains brings these amino acids into proximity; they are not adjacent—or even nearly so—in the sequence of amino acids that constitute the polypeptide chain. The "folding" of a molecule, incidentally, is determined by the sequence of amino acids of which it is made; the physical and chemical properties of the polypeptide chain, not an external factor (e.g., an enzyme), are responsible for its final shape.

Perhaps the ultimate example illustrating a function that accompanies the complex structure of proteins is provided by antibodies. The white blood cells that produce antibodies do not make antibodies that are designed especially for specific antigens. These cells, instead, possess the ability to recombine portions of the DNA sequences that encode the antibody protein chains. The outcome is that the antibody (a protein) borne on the surface of at least one of the hundreds of thousands of cells is likely to interact with any antigen and thus stimulate the proliferation of that cell and the synthesis of that particular antibody. The probability of an antibody-antigen reaction somewhere among the millions of differing antibodies is so great that recognition of one's own proteins ("self") and the subsequent destruction of those proteins (of one's own self) must be avoided. It appears that the function of the thymus gland just before birth is to screen white blood cells that have already reacted with antigens (those that normally must be "self" antigens) and to destroy them. The cells that remain in circulation, therefore, do not recognize the body's own proteins as antigens but are available to react with whatever foreign substances they may subsequently encounter.

Returning now to DNA, one may ask, "Where is the complex structure that would allow it to do more than store information?" The relation between its structure and its information-storing function need not be repeated in this chapter. James D. Watson and Francis H. C. Crick (1953) proposed a double-stranded structure for DNA in which the purine and pyrimidine bases of two antiparallel, complementary strands are held together by the two (i.e., A=T) or three (i.e., C≡G) hydrogen bonds that are formed by each pair of complementary bases. Information concerning protein structure as well as the time and place of protein synthesis resides in the long strand of chromosomal DNA. In addition, of course, DNA carries the information needed for self-replication.

The contrast between the complex three-dimensional structure needed for functions normally performed by proteins and the linear physical structure (much like the tape in a cassette) required by DNA for its function (information storage) suggests that, evolutionarily speaking, a still earlier chemical must have had both attributes: a complex structure for performing various tasks *and* a linear structure capable of accurate self-replication and information storage. From at least the early 1960s, attention has been focused on RNA as this early "living" substance; I recall a colloquium address given at Cornell University during that era by Francis Crick in which this role for RNA was his main thesis. (Joshua Lederberg [personal communication] has argued that anything RNA can do in the following respect, DNA can do just as well. Nevertheless, it may be wise to separate information bearing from an ability to *perform.*)

Tremendous support for the view that RNA is the substance that has both critical attributes was obtained by T. R. Cech and B. L. Bass (1986), who demonstrated that RNA may exhibit enzymatic properties. Previously, conventional wisdom held that all organic enzymes are proteins; the qualification "organic" is inserted to acknowledge the catalytic properties of many finely pulverized inorganic substances, including metals. Cech and Bass demonstrated that the unprocessed ribosomal RNA isolated from a protozoan and containing a large intron (a portion of the pre-rRNA molecule that must be excised before the rRNA can function in the construction of a ribosome) can splice out that intron in the absence of any protein. This pre-rRNA, that is, can process itself! And, indeed, the intron that is removed is the enzyme (*ribozyme,* to distinguish it from a proteinaceous enzyme) that brings about its own removal.

To function as an enzyme, the RNA molecule must be able to take on a complex three-dimensional shape. How it does so is evident in Figure 9-2. Long stretches of RNA must be able to undergo complementary pairing, an ability that requires inverted sequences of com-

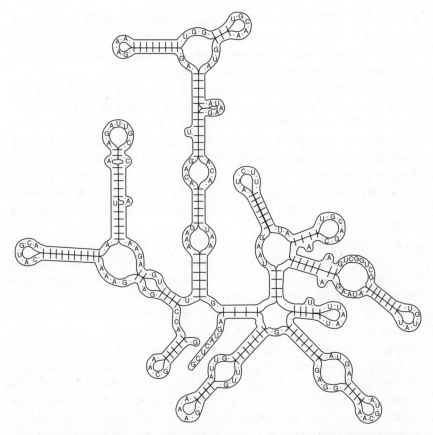

Figure 9-2. A molecule of ribosomal RNA (rRNA), whose function within the ribosome (which consists of three different RNAs and approximately 50 protein molecules) depends on its three-dimensional shape. In part, this shape is generated by many self-complementary foldback regions that pair to form double-stranded regions (the many A-U and C-G pairs are not specified in the figure). In addition, widely separated but complementary regions are also thought to pair. Readers may convince themselves that the DNA from which this rRNA molecule was transcribed must contain numerous palindromic regions (reverse repeats).

plementary purine and pyrimidine bases. Where this complementarity exists, RNA, as well as DNA, can form a duplex structure. Complementary regions that are remote from one another could generate a three-dimensional configuration.

Studies since the pioneering one by Cech and Bass have revealed that introns in many ribosomal RNAs, viruses, and mitochondrial and chloroplast rRNA and tRNA exhibit ribozymal function. Indeed, ribozymic introns appear to fall into two classes with somewhat different structures and different ways by which they carry out their function of processing of pre-rRNA and pre-mRNA.

Ribonucleic acid (RNA) possesses the essential characteristics of both DNA and proteins: it can direct the synthesis of a complementary strand (which, when it subsequently directs the synthesis of its complement, reestablishes the original molecule), and it can perform enzymatic functions, including those required for self-modification. For the purposes of the present text, establishing a foolproof account of a hypothesis concerning the origin of life is not necessary. Nevertheless, a generalized scheme such as the following can be outlined.

Any early RNA molecule that (because of its three-dimensional structure) interacted with any "organic" or inorganic particle possessing catalytic properties would in time become the predominant molecule. The catalytic property referred to here applies to the synthesis of complementary strands. The three-dimensional configuration of an RNA molecule is virtually identical to that of its complement (minus and plus strands) (Figure 9-3).

If the most efficient associated catalysts were short polypeptide chains, any RNA molecule that catalytically (enzymatically) enhanced the assembly of its associated polypeptide chain would, in turn, become the dominant RNA molecule (and its associated polypeptide chain would similarly become the predominant proto-protein).

In the absence of life as we know it today, degradation of complex molecules occurred primarily through physical processes (such as the thermal excitation of atoms) and, to a lesser extent, happenstance catalytic reactions associated with the chance encounters among molecules. Nevertheless, an RNA molecule that shifted (1) its information-related activities to the chemically more stable duplex molecule (DNA, today) and (2) its catalytic properties to associated polypeptide chains that were synthesized according to its own specifications would possess a tremendous Darwinian advantage. It would, that is, become the dominant DNA (RNA-DNA-protein complex) molecule of its era.

Whether the above account is true in any or all details is relatively unimportant. What is important is the implication that early DNA

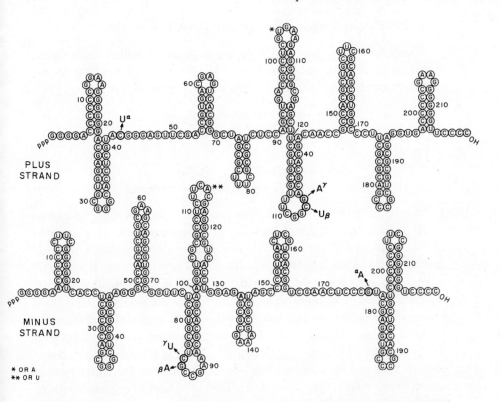

Figure 9-3. The structure of the residual portion of an RNA virus, Qβ (pronounced cue-beta), after approximately 100 or more "generations" of in vitro replication. All "nonessential" portions of the original phage, such as the portion that specifies coat proteins, have been lost through natural selection. Note that the two complementary single strands (plus and minus) form physically similar two- or three-dimensional structures; consequently, properties that depend upon the molecules' physical shape will be similar for both plus and minus strands. Incidentally, the internal foldback regions prevent the complementary genomic strands from forming a single, large double-stranded molecule. (From *Basic Population Genetics* by Bruce Wallace. Copyright © 1981 by Columbia University Press. Reprinted with permission of the publisher.)

molecules were constructed according to the base sequences of existing RNA molecules. The complex structure of these RNA molecules was achieved by means of numerous, internally homologous sequences of base pairs. Thus, the earliest DNA probably contained many reversely oriented sequences of bases (reverse repeats, or "palindromes").

Advantages accompanying repeated DNA homologies

Homologous regions of DNA are regions at which exchanges between DNA molecules occur most readily, perhaps because replication errors that lead to exchanges can occur in such regions:

```
>-----------G-C-C-T-A-G-C-A----------->
<--⌐r-----C-G-G-A-T-C-G-T<---⌐r----
   )(                    )(
<--⌐J L-----C-G-G-A-T-C-G-T<---J L----
---------->G-C-C-T-A-G-C-A----------->
```

Exchange of genetic material between DNA strands with homologous regions that differ somewhat in base sequences is a source of strand-to-strand variation. Such exchanges are the basis of and provide the Darwinian advantage for sexual reproduction. Evolution is speeded up by sexual reproduction; that is, by reproduction in which a single individual or cell comes to possess DNA from two sources and is able to produce progeny DNA containing combinations of the two (Figure 9-4). This process can be illustrated by the following three DNA molecules, in which the uppercase letters both individually and in combination are advantageous in a Darwinian sense. The intervening (vertical) lines represent regions within which recombination can occur:

$$\text{I.} \quad A \updownarrow b - c$$

$$\text{II.} \quad a \updownarrow B \updownarrow c$$

$$\text{III.} \quad a - b \updownarrow C$$

The putting together of the combination -A-B-C- can occur much more rapidly through recombination than by the rare chance mutation of two lowercase letters into uppercase ones in any of the three strands.

Internally homologous regions of DNA are useful not only with respect to sexual reproduction and the pace of evolutionary change but also, at least in rather complex organisms, with respect to coordinated gene action. Werner Maas (1961) described the synchronous repression and derepression of three genes in the arginine pathway in *Escherichia coli*. His model, like many that have been proposed

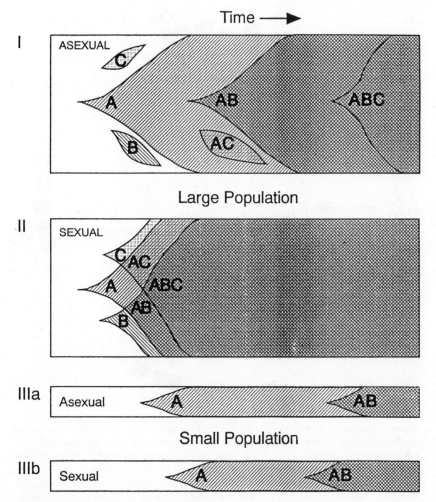

Figure 9-4. The role of sex in accelerating evolution. The initial population (left side) is assumed to consist of individuals whose genotypes can be written *abc*. The possession of any single "upper-case" allele enhances fitness, but allele *A* does so more than either *B* or *C*. In turn, the fitness of *AB* individuals exceeds that of *A* and *AC* individuals. Maximum fitness is exhibited by *ABC* individuals. Diagram I reveals that the genotype *ABC* can arise from *abc* individuals only by mutations that occur in a given sequence: *A* in *abc*, *B* in *Abc*, and *C* in *ABc*. In contrast to this sequential constraint, (II) sexual reproduction (recombination between DNA molecules of diverse origins), in large populations, allows *ABC* individuals to arise quickly by reshuffling appropriate DNA molecules. The lower diagrams (IIIa and IIIb) reveal that the selective advantage of sexual reproduction is manifest in large, not small, populations. (Adapted by permission of the University of Chicago Press and J. F. Crow, from Crow, 1965, *American Naturalist* 99:439–450.)

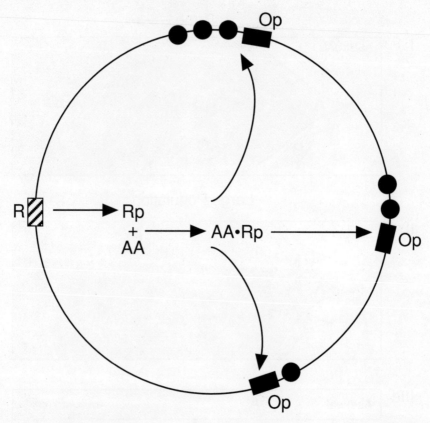

Figure 9-5. A model for gene control that leads to the coordinated operation of genes at three, widely separated loci on a bacterial chromosome. A repressor molecule, whose synthesis depends upon the *R* locus, combines with the amino acid whose synthesis is to be controlled; the repressor–amino acid complex (AA·Rp) then binds to a DNA sequence (Op, an operator) to prevent transcription of the gene encoding the enzymes for the synthesis of the amino acid itself. Because of the repressor molecule, the amino acid acquires the ability to terminate its own synthesis. Because of the identity, or near-identity, of the three chromosomal (operator) regions labeled Op, the circular chromosome can undergo "illegitimate" pairing and recombination that, if not prevented, could lead to the loss or rearrangement of the bacterium's DNA. (After Maas, 1961, by permission of Cold Spring Harbor Laboratory and W. K. Maas.)

since (see Britten and Davidson, 1969), involves a regulatory protein that, having interacted with an environmental signal (i.e., the amino acid arginine), attaches to the operator region of several, nonadjacent gene loci (Figure 9-5). The base sequences in these attachment regions of the bacterial chromosome have identical structures; they constitute internally homologous regions of DNA.

Risks accompanying repeated DNA homologies

Many useful purposes are served by the dispersed presence of homologous regions of DNA throughout a bacterial chromosome (or among the numerous chromosomes of higher organisms); three have been mentioned specifically: (1) the benefits that accompany sexual reproduction, (2) the synchronization of gene action, and (3) the provision for the three-dimensional structures of RNA molecules. These homologous regions are not risk free, however. By far the most common requirement for recombination is homology—the identity or near-identity of base-pair sequences. The term *homologous* refers as well to genetic map position; the *white, eosin,* and *apricot* alleles are homologous in this sense because they all occur at map position 1-1.5 (1.5 on chromosome 1) and in or near band 3C2 in Bridge's cytological map of the X chromosome. Recombination that occurs between identical base sequences (homologous in the first sense) that occupy different physical locations (nonhomologous in the second sense) can be disastrous for the organism. If the identical sequences are arranged in the same direction (abcd-XYZ-abcd), recombination excises a circular piece of DNA from the DNA strand; without a centromere, this portion would be lost in higher organisms. If, on the other hand, the identical sequences are arranged in reverse directions (abcd-XYZ-dcba), recombination leads to an inversion of the intervening segment that is rotated (is upside down) as well as inverted, end for end. Should the identical segments be located on separate chromosomes, translocations with the production of dicentric or acentric chromosomes are likely outcomes (Figure 9-6).

The problems posed for an organism by recombination initiated by sequence homologies within the individual's genome appear to be sufficiently serious as to require preventive measures. And, indeed, there do seem to be preventive mechanisms; meiosis in hexaploid wheat can serve as an example.

Triticum aestivum, an allohexaploid bread wheat, carries three diploid sets of chromosomes ($2n = 14$) that have been obtained through hybridization (followed by chromosome doubling) among three species of wheat. The latter will be referred to simply as species

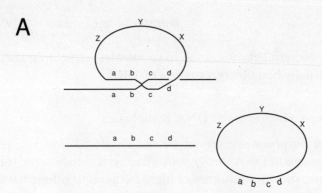

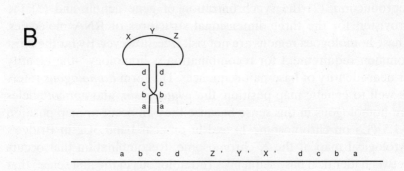

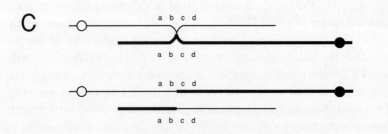

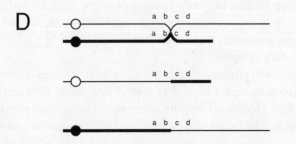

A, B, and D because even now there are differing opinions as to which of the many diploid species of *Triticum* are the "parental" species for this hexaploid ($2n = 42$).

Contrary to intuitive expectations, the fertility of polyploid species derived from the hybridization of near relatives is much poorer than that of polyploids derived from the hybridization of distantly related species. In the latter case, once the two haploid chromosome sets (say, *J* and *K*) of the hybrid have been doubled (*JJ* and *KK*), each chromosome has a homologous pairing partner (1_j1_j, 2_j2_j, ..., 1_k1_k, 2_k2_k, ...). In this case, the reduction of chromosome number during meiosis can occur with considerable regularity. Before chromosome doubling, when the newly formed (F_1) hybrid carried only one chromosome of each parental genome, the individual chromosomes had no pairing partners, and, as a result, meiosis was an exceedingly irregular process; extremely poor fertility (seed set) was the result.

When the parental species (A, B, and D in the case of hexaploid wheat) are reasonably closely related, considerable pairing may occur among comparable chromosomes ("homeologous," as opposed to "homologous") contributed by different "parental" species. Even after chromosome doubling, this similarity among chromosomes can lead to the formation of multivalent rather than the normal bivalent formation at meiotic metaphase. The inability of chromosomes to segregate in a regular fashion from trivalents considerably reduces the fertility of such hybrids. Unlike the higgly-piggly chromosomal pairing that might easily occur, *T. aestivum* exhibits standard bivalent metaphase figures during meiosis, and, consequently, its fertility is high, as it must be if it were to be a successful crop. These bivalents are formed by the pairing of homologous chromosomes only: 1_A1_A,

---

Figure 9-6. Diagrams illustrating the outcomes of the "illegitimate" pairing and recombination of identical segments of DNA. (A) The pairing of spatially separated tandem duplications and the excision of a circular piece of DNA; the linear chromosome loses the corresponding piece. (Note the reverse possibility: The circular piece of DNA can pair with the segment *abcd* of the linear piece and, by recombination, reinsert itself. This is a mechanism that leads to the incorporation of viral DNA into bacterial, plant, or animal chromosomes.) (B) The inversion of an intervening segment by recombination between inverted (but otherwise identical) regions of DNA. (C and D) The formation of dicentric, acentric, and translocated (but monocentric) chromosomes by the illegitimate pairing of identical regions that exist on nonhomologous chromosomes.

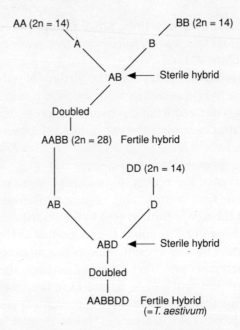

AA (2n = 14)   BB (2n = 14)

                    A   B

                         AB ◄─── Sterile hybrid

                Doubled
                    |
                AABB (2n = 28)   Fertile hybrid

                              DD (2n = 14)
                                   |
            AB   D

                         ABD ◄─── Sterile hybrid
                          |
                       Doubled
                          |
                       AABBDD   Fertile Hybrid
                              (= *T. aestivum*)

Figure 9-7. The origin of allohexaploid bread wheat (*Triticum aestivum*) through the sequential hybridizations of three species of wheat identified here merely as A, B, and D. Hybrids that carry only one copy of each chromosome of each species are largely sterile ("sterile hybrid") because the lack of a pairing partner prevents tetrad formation at meiotic metaphase and the subsequent orderly apportionment of chromosomes during anaphase. This difficulty is removed by the accidental doubling of chromosome number in germinal tissue (or its somatic predecessor). Thus, the fertile tetraploid possesses a *pair* of each chromosome from species A and B, whereas the hexaploid bread wheat has *pairs* of chromosomes of all three species.

1_B1_B, 1_D1_D, 2_A2_A, 2_B2_B, . . . , 7_A7_A, 7_B7_B, and 7_D7_D. In brief, this hexaploid wheat behaves as if it were a diploid!

The diploid-like behavior of *T. aestivum* is dependent upon the presence of a particular allele of a single gene, *Ph* (*Pairing homeologous*), that is located on the fifth chromosome of species B (5_B). Presumably, this gene was responsible for the success of the initial A × B hybridization (see Figure 9-7) and for the continued success of the subsequent hybridization (A × B) × D. Remove gene *Ph* from the hexaploid genome, and mispairing among homeologous chromosomes becomes rampant. The number of fertile grains of wheat that are set by *T. aestivum* lacking *Ph* is exceedingly small. We see evidence in wheat that comparable sequences of base pairs in chromosomal DNA can, if given the opportunity, mispair (illegitimate or homeologous pairing), with at times disastrous results. In forestalling the misfortunes that accompany abnormal recombination, genetic systems have evolved mechanisms that virtually guarantee that only legitimate recombination will occur—despite the ample opportunities to do otherwise.

Extrachromosomal elements

Our discussion of transposable elements required, as Erich Kästner asserted, that we best begin with entirely other things—notably, the probable origin of life as we know it. By the time DNA was involved, it was ordained to possess sequences of purine and pyrimidine bases that, if allowed to recombine freely, could lead to drastic physical rearrangements of the DNA molecule—including the loss of certain intervening segments. Indeed, one can imagine that primitive microorganisms, because they lacked a proper means for preventing rampant DNA recombination, generated extrachromosomal fragments in enormous numbers through the excision of small, circular DNA fragments (p. 193).

Modern bacteria are highly sophisticated organisms; that is a fact that should not be forgotten. They have undergone a natural selection for efficiency and successful existence that is coextensive with that of human beings or any other form of life. Consequently, we can regard neither bacteria nor the extrachromosomal elements (insertion sequences, plasmids, and viruses) they contain as primitive. The lambda phage, of which we shall speak below, is an extremely efficient, self-replicating device. At some time in the past, cast-off circular pieces of DNA happened to contain information that aided their continued existence, their survival. Because these primitive elements arose as the result of chance happenings, they must have been simple. "Simple" does not mean small and streamlined; it more likely means crude and unnecessarily cumbersome. Crude and cumbersome, however, in a more nearly barren and life-free world.

Of the transposable, autonomously self-replicating elements known today, the insertion sequences (IS) are the simplest (Table 9-1). IS elements are *defined* as transposable elements possessing no ability other than that required for excision from and insertion into

Table 9-1. Some insertion sequence (IS) elements found in bacteria, their length, and the nature of their inverted terminal repeats (number of identical bases of the two ends / total number of bases at each end).

Element	Occurrence	Length (base pairs)	Terminal repeats
IS1a	*Escherichia coli*	768	20/23
IS1b	*E. coli*	768	20/23
IS1c	*Shigella dysenteriae*	768	18/23
ISvE	*S. dysenteriae*	766	25/31
IS2	*E. coli*	1327	32/41
IS4	*E. coli*	1426	16/18
IS5a	*E. coli*	1195	15/16
IS5b	*E. coli*	1195	15/16
IS59R	phage	2534	8/9
IS101	phage	201	31/37
IS102	phage	1057	18/18
IS903	transposon	1057	18/18
IS10R	transposon	1329	17/22

bacterial chromosomes. That ability depends upon the presence of an enzyme, a *transposase,* which is encoded by a gene contained within the IS DNA. The ability to leave and enter the bacterial chromosome resides, apparently, in identical or near-identical terminal sequences of DNA that are at both ends of the IS unit and that are oriented in opposite directions. Segments of DNA whose two strands have the same sequence of base pairs when read in opposite direction

$$\rightarrow A \quad T \quad T \quad C \quad G \quad \quad C \quad G \quad A \quad A \quad T \rightarrow$$
$$\leftarrow T \quad A \quad A \quad G \quad C \quad \quad G \quad C \quad T \quad T \quad A \leftarrow$$

are called "palindromic" ("Able was I ere I saw Elba"). The structural gene within an IS unit converts it into what may be called a split palindrome ("Able was I for many years before I saw Elba"). Among the many IS units known, one consists only of the two inversely repeated terminal sequences of DNA. This simple unit is most likely an advanced, streamlined element that functions by virtue of the transposase that is synthesized under the direction of other, larger IS elements within the same bacterium.

Better known than insertion sequences are transposons; better known because of the way they are defined: A transposon possesses

Table 9-2. Some known transposons (Tn) in bacteria and the special characters not related to transposition that cause them to be called transposons rather than insertion elements (IS).

Designation	Size (base pairs)	Termini (base pairs)	Chromosomal target (base pairs)	Open reading frame(s)	Special character(s)
Tn3	4957	38 ITR	5	3	Ap^R
Tn5	5818	IS50 ITR	9	2	Km^R, Sm^R, Bl^R
Tn10	9300	IS10 ITR	9	1	Tc^R
Tn551	5300	40 ITR	5	ND	Em^R
Tn916	16,400	imperfect	none	ND	Tc^R
Tn4001	4700	IS256 ITR	ND	ND	Gm^R, Tm^R, Km^R

Note: Ap^R, ampicillin resistance; Bl^R, bleomycin resistance; Em^R, erythromycin resistance; Gm^R, gentamycin resistance; ITR, inverted terminal repeat; Km^R, kanamycin resistance; ND, not determined; Sm^R, streptomycin resistance; Tc^R, tetracycline resistance; Tm^R, tobramycin resistance.

one or more characteristics other than that required for excision from and insertion into a bacterial chromosome. The extra characteristic(s) provides the basis for designing selective techniques and thus revealing the transposon's existence. Table 9-2 reveals that many transposons have the ability to render their bacterial carriers resistant to antibiotics or heavy metals; these particular abilities may reflect, of course, the types of selective measures used in isolating and studying these elements. A comparison of Tables 9-1 and 9-2 clearly reveals that the extra properties exhibited by transposons require extra DNA relative to that found in IS units; the extra DNA carries the information needed for the synthesis of the enzymes responsible for the many resistances these transposons confer on their carriers.

Lambda phage

From the early beginnings of extra chromosomal elements—elements that survive at times by helping their microbial hosts to survive (see Table 9-2)—selection has produced some marvelously sophisticated viruses and phage particles. (For an excellent account of the lambda phage see *A Genetic Switch: Gene Control and Phage λ* by Mark Ptashne [1992].) Far from merely paying its own way within a bacterium by providing resistance to toxic or bacteriocidal substances, lambda phage senses the environment and largely determines its own fate—either one or the other of two that are open to it.

Lambda phage consists of a protein coat that surrounds a DNA chromosome consisting of approximately 50,000 base pairs. The coat consists of 15 types of protein, each of which is specified by a structural gene within the phage chromosome. After the phage DNA has been injected into a bacterium, the DNA may replicate many times, in which case protein coats are synthesized and an enzyme is produced that causes the destruction of the host cell and the release of 100 or more progeny phage (lysis). On the other hand, the phage DNA may become incorporated within the bacterium's chromosome and, as an inconspicuous portion of that chromosome, replicate in conjunction with that chromosome and under the control of the bacterium's replication machinery.

How is the "decision" to lyse (phage) or not to lyse (integrate) the host bacterium made? The account given here will follow Ptashne's. Once lambda DNA has been injected into a bacterium, lambda's fate rests with the bacterium's enzymes. The phage DNA that enters the bacterium as a linear molecule becomes a circle when complementary single-strand projections at the ends of the linear molecule pair spontaneously and the DNA is "repaired" by a bacterial ligase:

* GGGCGGCGACCT
CCCGCCGCTGGA *

where the asterisks represent points for ligation. The joining of these cohesive ends (COS in Figure 9-8) creates the circle.

Bacterial RNA polymerase is also responsible for the next step in the sequence following infection; the host polymerase transcribes the *N* and *cro* genes (short arrows at top of Figure 9-8) into mRNA, which is then translated into protein—once more by the bacterium's cellular machinery.

The N protein is a positive regulator protein that opens the genes to the left of *N* and to the right of *cro* to transcription. The *N* protein causes RNA polymerase molecules to ignore certain "stop" signals; as a result, mRNA molecules elongate to include several genes to the left of *N* and to the right of *cro*.

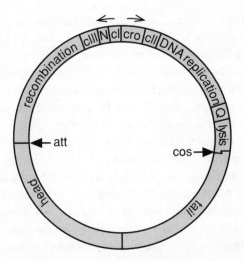

Figure 9-8. The chromosome of phage λ (lambda), showing the general location of genes that are responsible for recombination, head proteins, tail proteins, and phage replication. An early step following the infection of a bacterium by phage λ is the transcription of the *N* and *cro* genes as indicated by the small arrows. After that, a cascade of events determines, among other things, whether the phage DNA will be inserted into the bacterium's chromosome (lysogeny) or whether a large number of progeny particles will be formed and the bacterium destroyed (lysis). These decisions are discussed in the text. *att,* the attachment site at which the phage DNA "pairs" with the bacterium's chromosomal DNA; *cos,* the paired cohesive ends of the phage's linear chromosome (following injection into the bacterial cell), which, by pairing, create a circular phage chromosome. (After Ptashne, 1986.)

The decision that follows involves a choice between two pathways: for the cell to be lysed, head, tail, and *R* genes must be transcribed. For the cell not to be lysed but, rather, for the phage DNA to be inserted into the bacterial chromosome, genes *cI* and those required for chromosome integration by means of recombination (see Figure 9-8) must be transcribed. Here, the lambda phage senses its environment: the decision rests with the level of bacterial proteases and cAMP in the infected cell. The phage protein cII is the key: high levels of cII lead to lysogeny (integration); low levels to lysis. High levels of bacterial proteases lead to low levels of cII and, hence, to lysis. Low levels of proteases leave high levels of cII protein and lead to lysogeny. Active growth of bacteria—in complete medium, for example—leads to high levels of proteases and, hence, to the lysis of infected bacteria. Starved cells have little protease activity, thus leaving cII proteins to accumulate; the infection of starved cells by lambda phage usually leads to the incorporation of phage DNA into the bacterial chromosome. One is tempted to interpret the logic of this decision (what biologists refer to as a "just-so story"): Why attempt to create progeny phage in an inefficient, worn-out factory? And, for what purpose would they be created if the entire bacterial colony is similarly starved? Better to hunker down and wait.

The present chapter, "Transposable Elements," has been prepared to illustrate the role these elements have played in the study of gene action—the stated purpose of this entire text. Clearly, the study of the genes carried by the transposable elements themselves has shed considerable light on the mechanisms of gene action. Understanding the decision between lysis and lysogeny that follows the infection of a bacterium by lambda phage, allowed us to gain an understanding concerning the expression of genes lying in the northwest arc of the phage chromosome (Figure 9-8) and those occupying the remaining three-quarters of the chromosome. The proteases present in the bacterial cell at the moment of infection triggered the lyse–do not lyse decision. Subsequently, however, there is a cascade of decisions, each made on the basis of molecules generated by the preceding one.

Two phenomena of the many that control the "life" cycle of lambda phage deserve discussion. First, a bacterium whose chromosome contains the DNA of a lysogenic lambda phage cannot be infected by other lambda phages. Why? Examination of Figure 9-8 reveals the presence of a gene, cI, at the top of the chromosome. Early transcription begins on either side of that gene and proceeds both to the left and right. Simultaneously, cI is transcribed. In those cases where cII and cAMP concentrations have led to sufficient levels of cI production, the phage chromosome is integrated into that of the infected bacterium. The product of cI is a protein containing 236 amino acids, the repressor protein. It binds to the incorporated phage DNA and suppresses the transcription of all phage genes, thus ensuring that the phage remains quiescent in its new home. Excess repressor molecules bind to any new lambda phage DNA that might be injected into the bacterial cell by an outside infective phage particle. The result is that the new infective phage DNA is inactivated immediately upon entering the lysogenized cell.

The second phenomenon to be explained is the origin of the vastly different quantities of two proteins, we can call them A and B, that are made from a single messenger RNA that has been transcribed from both structural genes (polycistronic mRNA). Actually, there are several possible answers to this problem, including the mere probability that the ribosome by which translation of the mRNA occurs will become detached from the polycistronic mRNA. If this detachment is a frequent event, the leading gene will be completely transcribed more often than the succeeding one.

Lambda phage has a second means for achieving the same disparity in quantities of A and B. The polycistronic mRNA has a symmetrical complementary sequence of bases that, when the mRNA is transcribed, folds back on itself to form a double-stranded RNA. This portion of the RNA is attacked by a bacterial double-stranded RNase (dsRNase) and is destroyed. Once more, the leading gene will be available for translation longer than its succeeding companion. This pattern of control has been named: retroregulation.

Because viruses, plasmids, and other transposable elements contain regions that exert control over gene action, they have proved to be especially useful in the study of gene action in general. That these transposable elements—transposons, especially (see Table 9-2)—carry genes conferring resistance to various antibiotics is useful as well. By using restriction enzymes that attack known small palindromic regions of any DNA in which they occur, molecular geneticists are able to cut (restriction enzyme) and paste (reannealing of protruding complementary single strands) DNA from many sources into virtually any composition they desire. To study the activity of any gene whose activity is under metabolic control, one merely attaches to its normal control region a structural gene (reporter gene) whose product is easily assayed. The *lacZ* gene from *E. coli* is useful because its product, β-galactosidase, is easily detected and its quantity measured. Even better, for many studies, is the structural gene for luciferase, the enzyme that produces light by oxidizing luciferin: the intensity of light reveals the rate at which luciferin is being oxidized and, therefore, the quantity of enzyme. In addition, luciferases that produce light of different wavelengths exist; their structural genes have been "cloned" and are available for studies that reveal the temporal patterns of activity of two or more genes.

Transposable elements in higher organisms

The earliest convincing studies showing that genes (that is, chromosomal material with particular phenotypic effects) occasionally move spontaneously from one genetic map location to another (on the same chromosome or on a different one) were carried out by the late Barbara McClintock at the Department of Genetics of the Carnegie Institution of Washington, Cold Spring Harbor, New York. This reference is given in considerable detail because the Biological Laboratory at Cold Spring Harbor is the host to an annual prestigious symposium and is the laboratory at which geneticists from all continents have traditionally convened to carry out summer research. For more than half a century, Cold Spring Harbor has been the leading site for research in modern genetics—both research and the exchange of ideas.

Indian corn, maize or *Zea mays,* was McClintock's experimental organism. Beginning as a student in the late 1920s, she carried out both cytological and genetic research on corn. In 1931, she and her student colleague Harriet Creighton provided the first experimental proof that genetic recombination is accompanied by an exchange of chromosomal material as well (Creighton and McClintock, 1931). Later, using a small ring chromosome that had arisen (with half the original centromere) as a deletion from one of corn's ten chromosomes, McClintock was able to "dissect" a gene that controls pigmentation in the plant's leaves. In this study, she relied on the frequent formation of double-sized dicentric rings following chromosome duplication and cell division. These large rings break during anaphase; if the break occurred in the *brown* locus, the "incomplete" gene resulted in a streak of abnormal color in the developing leaf.

Another of McClintock's studies involved the fusion of the newly created "broken" ends of chromosomes. When a broken chromosome duplicates in preparation for cell division within the endosperm of the corn kernel (not in the embryo), the two resulting chromatids "fuse" at their free ends, thus producing a dicentric daughter chromosome that, having itself been forced to break during the next cell division, undergoes "fusion" once more as it prepares for the next cell division: the so-called breakage-fusion-bridge cycle. The result of this much-repeated cycle, shown in Figure 9-9, is a sequence of repeated, inversely oriented DNA segments. For reasons that are not clear even

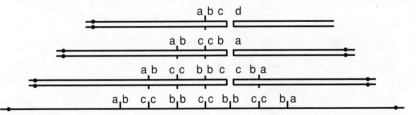

Figure 9-9. An illustration of the generation of duplicated reverse repeats by the chromatid breakage-fusion-bridge cycle in maize. The dicentric that is formed by the topmost fusion is symmetrical about c. The next dicentric is symmetrical about its fusion point and, therefore, is symmetrical for previously existing reverse repeats. The extra chromatin gained by the larger of the dicentrics consists, for the most part, of sequences of reverse repeats, although not necessarily as regularly spaced as those illustrated.

now, the breakage-fusion-bridge cycle induces genetic instability in corn. Nina Fedoroff (1984, 1991) emphasized the emergence of genetic instability under stressful conditions; however, the virtually mass production of repeated reverse repeats in the affected DNA (a pattern now known to be able to generate triple-stranded DNA) should not be ignored.

The account of transposable elements presented here will deal with the Activator-Dissociation system, which, historically, was the first system whose inheritance was understood. Note that a two-gene (dihybrid) system is much harder to resolve than a single (mono-hybrid) system. In addition, McClintock was confronted with variable phenotypic expressions (Dissociation, *Ds*, may cause chromosomal losses during cell division or, at a different physical location, gene mutations) and with the hitherto unheard of ability of genes to transpose—i.e., to alter their physical position within the genome.

Figure 9-10 illustrates the sorts of corn kernels available to McClintock for analysis; these and kernels with similar but not identical patterns would be scattered throughout ears of mature corn. First, as in 9-10A, there were the expected, full-colored kernels. Then, there were those (Figure 9-10B) that were uniformly colorless. There were also colorless kernels on which were scattered variously sized spots of color. And, as shown in Figure 9-10D, there were kernels that were mottled colored and colorless. The last were the first to be understood: the colorless regions were those in which a terminal portion of chromosome 9 carrying *C* (aleurone color) had been lost,

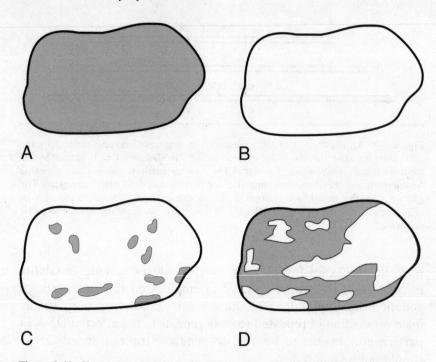

Figure 9-10. Sketches of four kernels of Indian corn (maize, *Zea mays*), illustrating the interactions of the Activator (*Ac*) element, the Dissociation (*Ds*) element, and the color gene (*C*) that is responsible for the red pigment in the kernel's aleurone layer of cells. (A) A solidly colored kernel illustrating the effect of *C*. (B) A colorless (white) kernel resulting from the inactivation of *C* by the nearby insertion of a *Ds* element. (The phenotypic expression of a *Ds*-inactivated *C* locus mimics that of any one of many *c*—colorless—alleles.) (C) The restoration of the ability to synthesize pigment in patches of cells within which *Ds* has been removed from *C* by the action of the Activator (*Ac*) element. (D) The patches of pigmented and nonpigmented areas caused by the *Ds*-mediated loss of chromosomal material (and, hence, the loss of *C* in *C/c* heterozygotes); *Ds* mediates this loss only in the presence of *Ac*. (Drawn from photos in Fedoroff, 1984.)

thus leaving the cell with only *c* (colorless aleurone). McClintock could demonstrate that genes other than *C* had been lost together with *C*. The loss of *Wx* (waxy endosperm) and the concomitant expression of its recessive allele (*wx*) on the homologous (and retained) chromosome arm could be revealed by staining the endosperm with an iodine solution. (The heavy boundaries surrounding the colorless areas in Figure 9-10D are genuine; presumably an unused substrate that accumulates in the colorless sectors diffuses a short distance into the colored areas, adding to the intensity of pigment in these bordering cells.)

The loss of the terminal portion of chromosome 9 depended upon the presence of an "element" (McClintock avoided the term *gene* because it had acquired different meanings in the minds of different persons, thus rendering discussions difficult or even impossible) at the point of breakage; most chromosomes lack this element. The loss also depended, however, upon the presence of a second element—Activator, or *Ac*—somewhere in the cell's genome. With *Ac*, the chromosome broke at the site of *Ds;* without *Ac,* the chromosome was stable. The pattern of loss changed with the actual number of *Ac* elements: 1 *Ac,* large early blotches of colorless aleurone; 2 *Acs,* smaller blotches (somewhat as illustrated in Figure 9-10C); and 3 *Acs,* very small but numerous blotches that appeared, as McClintock once described them, as if they were to occur in the next cell division—one that did not take place.

The movement of these elements from one chromosomal site to another was perhaps the most startling observation, one that created many skeptics (but see Table 5-3 for irrefutable evidence). McClintock found, however, that movement was an asymmetrical property: *Ac* could move about with or without *Ds* in the cell's genome; *Ds* could move, however, only in the presence of *Ac*. This asymmetry led to the conclusion that *Ds* was a defective *Ac* element.

The molecular work on maize controlling elements has been carried out primarily by Nina Fedoroff and her colleagues. Their research has revealed the physical bases for McClintock's early observations: *Ac* is an element about 4500 bases long that encompasses two structural genes and whose ends consist of inverted repeats:

$$\rightarrow \text{TAGGGATGAAA} \ldots \ldots \text{TTTCATCCCTG} \rightarrow$$
$$\text{two genes}$$
$$\leftarrow \text{ATCCCTACTTT} \ldots \ldots \text{AAAGTAGGGAC} \leftarrow$$

One of the two *Ac* genes codes for a transposase, presumably one that recognizes and manipulates the terminal sequences.

A variety of *Ds* elements have been identified, three are shown diagrammatically in Figure 9-11. Each of the three either totally lacks the structural gene for transposase or carries an obviously defective

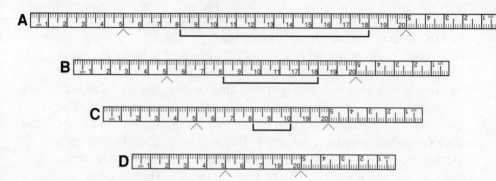

Figure 9-11. Schematic diagrams depicting the physical structures of (A) *Ac* (Activator element) and (B, C, and D) three *Ds*'s (Dissociation elements) in maize. Because *Ac* must include a functional transposase gene (as well as one other gene), its structure is a given. (In diagram A, the bar that extends from 8 to 18 represents the functional transposase gene.) The inverted terminal regions (see text) are represented as 1–5 and 5–1. Because the two strands of DNA are antiparallel, an inverted DNA segment must also be turned upside down (see p. 193.) *Ds* elements, since they are defective *Ac* elements, can have a variety of structures becuse the transposase gene can be inactivated in many ways. Diagrams B and C suggest that portions of the transposase gene have been deleted; in D, the gene has been entirely deleted.

transposase gene. The enzyme transposase can cause the excision and transposition of any DNA segment possessing inverted terminal repeats essentially identical to those of the *Ac* elements. Because they lack the ability to synthesize transposase, *Ds* elements remain fixed in position unless an *Ac* element is present in the cell's genome. One can, of course, imagine a *Ds* element with a defective terminal repeat (a piece of DNA with a fixed position in the genome; one that could be interpreted as a pseudogene if the element's defect in the transposase gene were slight), or an *Ac* element also with defective terminal repeats (a "fixed" gene causing instability in the physical structure of chromosomes).

In concluding this account of transposable elements in higher organisms, let us briefly survey those that have been identified in *Drosophila melanogaster* populations (Figure 9-12): for example, P elements, FB elements, *copia*-like elements, and still others.

The *P* elements of *D. melanogaster* were discovered during the mid-1970s. The initial observation involved "hybrid dysgenesis": flies obtained by matings of male and female flies of separate geographic origins sometimes exhibit a number of abnormalities involving steril-

P elements

Copia -like elements

FB elements

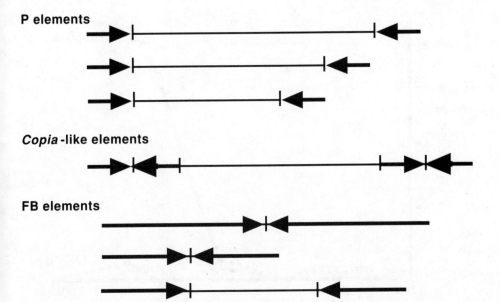

Figure 9-12. Diagrams illustrating the structures of several of the more common transposable genetic elements in *Drosophila melanogaster*. DNA segments of reverse orientation ("palindromes") are represented by arrows that point in opposite directions. Both P and FB elements have terminal regions that are palindromic. *Copia*-like elements have terminal regions that are themselves palindromic.

ity, male recombination (*D. melanogaster* males generally having no crossing-over), elevated mutation rates, and chromosomal breakage and recombination (for a review see Bregliano and Kidwell, 1983). The effects are generally asymmetrical in the sense that the outcomes of reciprocal crosses differ (Figure 9-13). That not all interpopulation crosses lead to dysgenesis should be emphasized; most such crosses result in progeny that exhibit heterosis (hybrid vigor).

The asymmetrical results have been resolved. There exist two kinds of flies with respect to P elements: *M* strains and *P* strains. Progeny obtained by crosses among either *M* strains or *P* strains exhibit no dysgenesis. Nor do the progeny obtained by mating *M* males and *P* females. Only the cross *P* male × *M* female results in dysgenic offspring, and that in perhaps only certain environments (Figure 9-13). Molecular studies involving the isolation of P elements and in situ hybridization of giant salivary gland chromosomes reveal that *P* strains of flies (but not *M*) have these elements scattered throughout their genomes, perhaps as many as 30–50 copies.

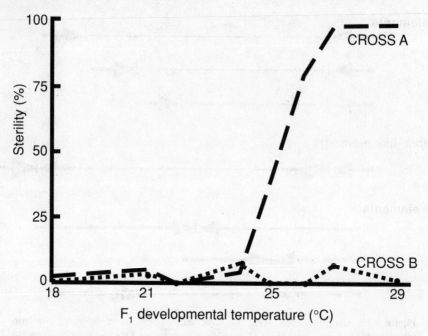

Figure 9-13. A diagram illustrating the striking difference in the degree of dysgenesis (represented here by sterility) exhibited by hybrid *Drosophila melanogaster* obtained by reciprocal crosses of two strains of flies. The P elements carried by males of one strain induce virtually complete sterility in offspring obtained by crosses (A) with females lacking these elements. No comparable effect is observed among progeny obtained by the reciprocal cross (B). The diagram also reveals that, like numerous other genetic traits, dysgenesis is strongly affected by the temperature at which the progeny flies are raised. (After Kidwell and Novy, 1979, by permission of *Genetics* and M. G. Kidwell.)

Understandably, the use of P factors in the study of gene action promises much more than has been gained so far. The insertion of these (and other transposable) elements into either structural genes or the control regions of structural genes results in gene mutation. Indeed, there is reason to believe that most naturally occurring mutations in *Drosophila* owe their existence to the insertion of transposable elements into one or the other of thousands of gene loci rather than to point mutations that alter single base pairs (and single amino acids).

An interesting speculation concerning the regulation of P elements within flies of P-bearing strains touches on the matter of gene action.

Recall that progeny produced by matings of male and female flies of P strains show no dysgenesis. The inactivation of these elements by P strains may involve a repressor of transposition or, possibly, an erroneous transcription of the transposon gene of this element. The faulty transposon may then "poison" either the functional enzyme, by forming an inactive complex with it, or the site at which the functional enzyme would normally act (Moran and Torkamanzehi, 1990).

Copia, copia-like, and FB (fold back) elements are also transposable elements that are capable, upon insertion within or near a functional gene, of inhibiting or modifying gene action. Thus, they result in gene mutations—mutations in the sense that the nearby gene no longer functions properly. As in the case of P elements, however, the entire arena of the developmental genetics of *Drosophila* (and other higher organisms) is so large and complex that it dwarfs whatever feeling we have of the known genetic effects of these (and other) elements. In brief, much of what is known is still in the descriptive stage—even descriptive in a molecular sense—so that the role these elements will play in the study of gene action largely remains to be seen.

10 Tailoring Genes

In the introductory comments (Chapter 1), we described the search for the gene as an ever-narrowing, linear search through time. The search progressed by steadily reducing the possible sites at which the gene might reside and then its possible chemical composition until at last the gene was found. Sherlock Holmes expounded a famous dictum: when you have eliminated all possibilities but one, then that one must be the correct one. Perhaps this dictum justifies a policeman's firing at a full-length drapery in a seemingly empty room. The search for the gene did not end quite that simply. The gene did not turn out to be (behind) the drapery (which was proteinaceous) but instead, it proved to be what was generally regarded as the curtain's supporting rod (DNA). The rather rigid, apparently repetitive, fibrous DNA proved to be the stuff of which genes are made; it did not merely provide the physical support upon which the gene rested.

The study of gene action, unlike the search for the gene, has been nonlinear, and in Chapter 1 we proposed a harp as an appropriate analogy (Figure 1-1). The harp strings become successively shorter (i.e., younger) through time. If each string represents a method enabling one to obtain at least some insight as to how genes perform their functions, then the longest string can be said to represent morphological observations. That long string has been examined and reexamined by the use of more and more recent techniques, including the most modern.

Color variation is also represented in our harp by one of the longer (i.e., older) strings. It served as a guide to Mendel in his initial work on peas, and, by virtue of the colorfulness of flowers, it provided plant physiologists with splendid material for studying gene action. Perhaps not unexpectedly, the plethora of plant pigments seduced most investigators into studies of the plant pigments themselves rather than into the genetics of plant pigments. An intriguing advance (Wood et al., 1989) has been made: luciferase genes of a click beetle, *Pyrophorus plagiophthalamus,* have been isolated and cloned in *Escherichia coli.* These genes, all of which make enzymes that oxidize luciferin, produce colors of different wavelengths. One obvious question concerns the chemistry of luciferin oxidation; that is a question for chemists. For the geneticist interested in gene action, these luciferases present an entirely different opportunity. When the beetle's structural gene is attached to the promoter region of any other plant of animal's gene, the synthesis of luciferase in the presence of luciferin provides an instantaneous record (i.e., report) of that gene's time of action. Different colors, in fact, enable the investigator to study several genes (homologous or nonhomologous) simultaneously.

In the present chapter, we reach the newest, shortest strings of our imaginary harp. The strings of this region represent questions that have arisen only because of the most recently discovered techniques. One could not ask about the role of repetitive DNA until one had discovered that repetitive DNA exists. Likewise, one cannot ask about the role of each base pair in a six-base-pair segment of DNA until one has techniques for identifying that short segment and then for substituting an alternative base at each position. Nor are six substitutions enough! The possible higher-order interactions among the bases in a segment even this short requires that substitutions be made 2, 3, or even 4 and 5 at a time. A thorough study requires not merely the three possible alternative substitutions at each of six sites (18 tests) but also an enormously large study of which the first step is the 9 possible alternative substitutions at each of 20 possible pairs of sites (180 tests). This introduction, then, sets the stage for the sections that follow: The study of gene action is an enormous, and enormously complex, undertaking.

Reliance on chance events

A rare event may be defined as one for which the probability of its occurring is extremely small. That definition, however, does not preclude its actual occurrence because given sufficient opportunities the probability of occurrence may approach 100%. Thus, a microbial geneticist searching for a streptomycin-resistant strain of *Escherichia coli* may face odds as low as 10^{-9} that a resistant mutant individual will arise by chance. Nevertheless, if 10^{10} sensitive individuals are distributed on a petri dish, 10 resistant individuals should be among them. Finding these rare individuals is not a difficult problem, because 10^{10} cells can be easily grown and spread on medium in a petri dish. If the culture medium contains streptomycin, only those 10 resistant individuals will form bacterial colonies. Although 10 is the expected number of colonies, the actual number will vary from 4 or 5 to 14 or 15. The probability of seeing no resistant colonies on this plate, small as the number 10^{-9}, may seem, is extremely small itself: 5×10^{-5}.

Many facets of life depend upon the near certainty that some rare event will occur. Woodpeckers dine on the insects that live under the bark of dead trees. Trees are long-lived organisms. Nevertheless, within a forest there are many dead trees, and, therefore, woodpeckers can thrive. (Conversely, deforestation may lead to the extinction of the resident woodpeckers long before all the trees have been hewn.)

Both Lamarck and Darwin were evolutionists. Lamarckian evolution relied upon the transmission of information from the environment to the organism and upon the development of an appropriate response on the part of the organism. The classroom example of giraffes stretching their necks to obtain leaves from the still-ungrazed upper branches of trees may be simplistic but it is effective, nevertheless. In the example concerning streptomycin resistance in bacteria, the Lamarckian view would hold that streptomycin "made" its presence known to the bacteria and the latter responded (with a 10^{-9} chance of success) appropriately by mutating to a resistant form.

Darwin's view of evolutionary change was fundamentally different from Lamarck's. Darwin postulated the preexistence of heritable variation and, in the face of environmental challenges, the survival and reproduction of some of these variants and the death or sterility of others. Thus, the store of existing heritable variation shifted in response to changing conditions. The resolution of these alternative explanations for the occurrence of streptomycin-resistant bacteria constitutes a classic study of the 1940s in microbial genetics. Both a statistical test devised by Salvadore Luria and Max Delbrück and a physical procedure used by Joshua Lederberg showed that streptomycin-resistant bacteria are to be found in bacterial cultures that have never encountered that antibiotic—they arise by chance and constitute the surviving portion of the culture when the latter is suddenly inundated with streptomycin (see p. 37).

This lengthy preamble has been provided in order to describe the mammalian immune system of higher vertebrates (including human beings), or at least that part of it responsible for antibody production. The observable fact is that when any of a large number of foreign substances is injected into an individual (mouse, rat, horse, or human being), the blood serum of that individual contains proteins (i.e., antibodies) that will interact with, agglutinate, and precipitate that foreign substance. Rarely, if ever, does such a reaction accompany the use of serum from a noninoculated individual; a strong reaction accompanies the use of serum only from an inoculated one.

The array of substances that can be tested for antibody formation is virtually unlimited. The antibodies that are found subsequent to the "alien" injection are proteins. Proteins are gene products; their amino acid sequences are specified by DNA. How can so many antibodies be made? The Lamarckian view would be that the injected substance elicits an appropriate response. Perhaps in the hundreds of millions of years that mammals and their ancestors have existed, every possible foreign substance has found its way into individuals' bodies and, by now, an appropriate gene awaits, ready to produce its antibody protein in response to a particular challenge.

Sir F. MacFarlane Burnett, Australian immunologist and 1960 Nobel Laureate, emphasized the mammalian genome's inability to harbor the multitude of genes needed to produce each of the hundreds of thousands of antibodies; the length of the genomic DNA simply is not great enough to meet such demands. Burnett postulated a mechanism by which somatic variation at one or a few loci could generate cells capable of producing many different types of antibodies. Those cells whose antibodies met corresponding antigens could then be amplified in number—a Darwinian selection model rather than a Lamarckian induced one. Within our (using human beings as the example) chromosomes are several regions that are responsible for specifying the amino acid sequence of antibodies. Within the precursor of those cells that actually produce antibodies (B lymphocytes), this region is destabilized: it undergoes deletions, inversions, transpositions, and a variety of splicings. These are changes in the DNA itself, unrelated to transcription (RNA synthesis) or translation (protein synthesis). The actual pattern of events differs from cell to cell among thousands of stem cells. As a result, perhaps as many as 100 million different antibodies are made—some are capable of reacting with this substance, others capable of reacting with others; some are capable of reacting with this portion (epitope) of a molecule; others with a different epitope of the same molecule. In any case, having encountered a foreign substance with which its unique antibody reacts, the cell is induced to divide and redivide, thus producing that antibody in great quantity. In essence, antibody formation is comparable to the growth of resistant bacterial colonies on antibiotic-laden media; it represents selection for amplification among cells carrying preexisting variant antibodies. (How individuals avoid self-destruction at the hands of their own immune system is an intriguing story as well [see p. 185]; some persons, in fact, are unable to do so.)

Having noted the success that has been achieved by the mammalian immune systems on the basis of random changes in the DNA structure of individual genes, biochemists (see Oliphant and Struhl, 1989) have investigated the consequences of replacing the active sites of enzymes with more or less random sequences of amino acids. The study by Arnold R. Oliphant and Kevin Struhl involved the enzyme, β-lactamase, that commonly confers on *Escherichia coli* resistance to penicillin and related substances (β-lactams). The active site for this enzyme lies within a 17–amino acid sequence (positions 61–77) of a polypeptide chain that is 286 amino acids long. The study was extremely large in scope: 100,000 altered enzyme molecules were independently generated; these were maintained in *E. coli* by appropriate vectors (plasmids; see Chapter 9). The sequences of purine and pyrimidine bases were determined for ten unselected clones merely to learn what the experimental procedure was generating. On average, 8.5 bases were found to be altered in every collection of 51 base pairs (17 amino acids); these might be expected to involve 6–7 of the 17 amino acids. This examination, in fact, revealed that neither the substitution of one base by another nor the site at which the substitution occurred was random. Of the 100,000 altered enzymes that were produced, about 2000 were able to protect their bacterial carriers against ampicillin. Of these, 58 were chosen for detailed study.

Table 10-1. The relative abilities of wild-type *Escherichia coli* and 10 strains carrying artificially generated variants of the β-lactamase gene to grow in the presence of 14 antibiotics (high numbers indicate high resistance) or to have their β-lactamase inactivated by heat (37°C). The artificially generated enzymes had 17 amino acids normally surrounding the active site replaced by randomly generated arrays of 17 amino acids. (After Oliphant and Struhl, 1989.)

	Bacterial strain										
	WT	1	2	3	4	5	6	7	8	9	10
Ampicillin	7	7	7	7	7	7	7	6	5	6	4
Piperacillin	7	7	7	6	7	6	7	4	4	6	0
Clavulanic acid	2	2	3	2	2	5	5	6	6	5	1
Sulbactam	2	1	2	2	2	5	3	2	2	5	4
Augmentin (mix)	5	5	5	4	5	7	6	7	6	6	1
Timentin (mix)	5	5	5	5	5	7	6	6	4	6	0
Cephaloridine	5	6	6	5	5	3	3	1	0	2	3
Cephalothin	5	7	5	4	4	2	2	0	0	1	1
Nitrocefin	5	3	5	4	3	1	1	0	0	1	0
Cefazolin	5	6	5	5	4	3	4	0	0	2	2
Cefaclor	5	7	7	7	7	7	4	0	0	0	2
Cefoperazone	5	6	6	6	6	7	5	4	3	4	3
Moxalactam	5	6	5	5	4	3	3	1	1	2	3
Cefonicid	5	6	5	5	4	3	2	0	0	2	1
Avg. of drug scores	5	6	5	5	5	4	3	2	1	2	2
Heat stability*	5	5	6	5	5	4	7	7	5	5	4

*0 = most heat labile.

They were tested for their ability to protect their carriers against 14 penicillin-like antibiotics (alone or in combination with a β-lactamase inhibitor) and for heat stability. The sequence of 17 amino acids in the active site (amino acids 61–77, inclusive) was also determined for each of the 58 altered enzymes. Tables 10-1 and 10-2 summarize these results, but for only 10 of the 58 altered enzymes described in the initial report.

Table 10-1 lists the relative activities of the wild-type and ten altered enzymes with respect to antibiotic and heat resistance. The second to the last line of the table records the average drug resistance of each of the 11 enzymes. Replacing one or more amino acids in the original wild-type enzyme oftentimes has unexpected results; namely, it enhances the enzyme's ability in one or more respects. For example, among the ten altered versions selected here (selected in part for just such reasons), six are more resistant to the inhibitory effect of clavu-

Table 10-2. The differences in amino acid composition of the wild-type *Escherichia coli* β-lactamase and the ten artificially generated enzymes whose relative activities were given in Table 10-1. It would appear from the larger number of artificial enzymes reported on in the original report (Oliphant and Struhl, 1989) that proline, serine, and lysine at positions 67, 70, and 73 are essential for enzyme activity.

Position	WT	Bacterial strain[†]									
		1	2	3	4	5	6	7	8	9	10
61	Arg	Thr	—	—	—	—	—	—	—	—	—
62	Pro	—	—	—	—	Ser	—	—	—	—	—
63	Glu	—	Asp	Asp	—	—	Asn	Psp	Lys	—	Gly
64	Glu	—	—	—	—	—	—	—	—	—	—
65	Arg	—	—	—	—	—	—	—	—	—	—
66	Phe	—	—	—	—	—	—	—	—	—	—
67*	Pro	—	—	—	—	—	—	—	—	—	—
68	Met	—	—	—	—	Val	Ile	—	—	—	—
69	Met	—	—	—	—	Leu	Leu	Ile	Ile	Leu	—
70*	Ser	—	—	—	—	—	—	—	—	—	—
71	Ser	—	—	—	—	—	—	—	—	—	—
72	Phe	—	—	—	—	—	—	—	—	—	Leu
73*	Lys	—	—	—	—	—	—	—	—	—	—
74	Val	—	—	—	—	—	Ile	—	—	—	—
75	Leu	—	—	Ile	—	—	—	—	—	Pro	—
76	Leu	—	—	—	—	—	—	—	—	—	—
77	Leu	—	—	—	—	—	—	—	—	—	—

*No amino acid substitutions were observed at positions 67, 70, and 73 among all 58 polypeptides analyzed.
[†]— indicates no substitution at the indicated position.

lanic acid than is the wild-type enzyme. Several exceed the wild-type enzyme in resistance to the two compounds Augmentin and Timentin (both are mixtures of a β-lactam antibiotic and the β-lactamase inhibitor clavulanic acid). Three of the altered enzymes are less labile at high temperature (37°C) than is the wild-type enzyme; four others are equally resistant to heat inactivation. Only one of the altered enzymes exhibits a higher average drug resistance than the wild-type enzyme, thus illustrating the well-known (but often forgotten) fact that the act of living demands numerous trade-offs; it does not require perfection.

Table 10-2 lists the actual amino acid substitutions exhibited by the ten altered enzymes discussed in Table 10-1. Eight of the 17 amino acids have been replaced in one or the other of these ten. In the larger sample of 58, only three amino acids remained invariant: proline at site 67, serine at site 70, and lysine at site 73. The absence of amino acid substitutions at these three locations suggests that they form the active site responsible for the hydrolysis of many β-lactam antibiotics, thus conferring resistance to the bacterium. In passing, it might be noted that penicillin is such a common feature of the bacterial environment that many resistant organisms are constitutive mutants within which 1%–2% of the total protein may be β-lactamase. The clavulanic acid mentioned in Table 10-1 is an inhibitor that binds tightly to the enzyme but is cleaved slowly, thus converting enzyme molecules into stable and inactive enzyme-substrate complexes.

Catalytic antibodies

This account of reliance on chance events may be concluded with a brief mention of antibodies once more—specifically, mention of *catalytic antibodies*. Emphasis in the earlier discussion centered on the random (or at least haphazard) construction of antibody proteins and (merely as a result of their enormous numbers) the high probability that *at least one* (often more) antibody will react with molecules of any foreign substance. The term *react* has proved to be an apt choice for this active, dynamic association which involves a true physical-chemical bonding that distorts both the antibody protein and its antigen (in effect, a substrate). The dynamic nature of this bonding is revealed, for example, by the finding that one particular essential amino acid in a proteinaceous epitope is normally concealed within the folded protein molecule. In order to be part of that particular epitope, the protein molecule must be unfolded so that particular amino acid becomes exposed to the antibody molecule.

The true nature of antibody-antigen interactions, and the concomitant transfer of energy between the interacting molecules (relaxing one, distorting the other), raises the possibility that such molecules may serve as catalysts (see Lerner and Tramontano, 1988). The first step in demonstrating that they might be is to postulate a structure for the transitory, high-energy molecule that may be formed while a substrate is disassociated by an enzyme (Figure 10-1). (Recall that an enzyme cannot carry out an impossible reaction; it can only accelerate one that otherwise occurs extremely slowly. The slowness results from the need for [activation] energy in creating the transient molecular state.) The second step is to find or create a molecule one

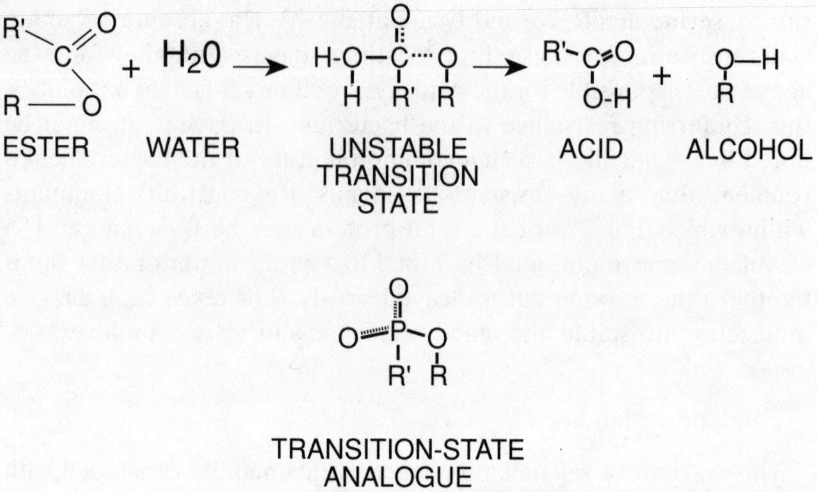

ESTER   WATER   UNSTABLE   ACID   ALCOHOL
TRANSITION
STATE

TRANSITION-STATE
ANALOGUE

Figure 10-1. (Top) The hydrolysis of an ester (left) and the resultant production of an acid and an alcohol (right). In the center is a postulated unstable transition state that exists briefly, following the attachment of water to the ester but preceding the formation of the end products. (Bottom) A stable analogue that resembles in many ways the structure of the postulated transition state. (After Lerner and Tramontano, 1988.)

portion of which is a stable region that mimics the physical aspect of the otherwise unstable, transient state. Such a stable analogue may require the substitution of one atom for another: phosphorus for carbon, for example (see Figure 10-1).

The artificially created molecule is now injected into a mouse or rabbit in order to induce antibody formation (Figure 10-2). Individual antibody-secreting cells are then "immortalized" by fusion with cancerous (myeloma) cells; otherwise, tissue cultures of the secreting cells would self-destruct after fifty or so cell generations. Each antibody-secreting cell produces an antibody that reacts with one or the other of the many epitopes characteristic of the injected molecule. Each *monoclonal* antibody can now be tested for catalytic activity, for the disassociation of the given substrate. Most will prove not to possess catalytic properties because they will attach to epitopes lying elsewhere on the injected molecule. One or more, however, may prove to have enzymatic properties—the goal of the whole exercise. Finally, one can demonstrate that the *catalytic antibody* does indeed react with the transient-state analogue of the injected molecule; that

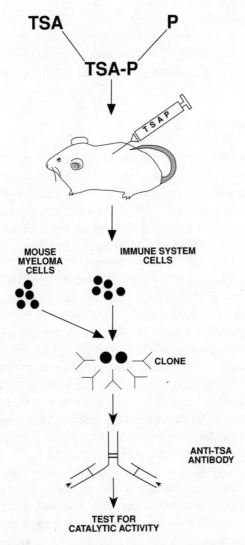

Figure 10-2. Obtaining an antibody that might react specifically with the postulated transition state of Figure 10-1. The transition state analogue (TSA) is attached to a carrier protein (P), and the complex molecule (TSA-P) is injected into a mouse (center). The mouse produces antibodies against the injected material; each of many cells possesses an antigen that reacts with one or another aspect (*epitope*) of the injected material. Each of these many antibodies is obtained in pure form by hybridizing individual immune system cells with individual mouse myeloma cells; each pair produces an "immortal" clone. The antibodies of each clone are tested against the analogue; those that react with the analogue are then tested for catalytic activity (for the conversion of an ester to an acid and alcohol, in the present case). (After Lerner and Tramontano, 1988.)

analogue is the antibody's epitope. Removing the analogue, one finds that the catalytic antibody no longer reacts with the injected molecule. In short, the epitope for the catalytic antibody is the transient-state analogue, not some other portion of the injected molecule.

Transgenic organisms

The bulk of all hereditary transmission on earth is accomplished by the replication of DNA and its passage from one cell (the parent) to two or more *progeny* cells. Replication is a strikingly conservative process: how else can the 4×10^{13} red blood cells normally present in one's body at a given moment all contain the identical hemoglobin molecules, of which there are millions per red blood cell? Even the highly variable HIV virus that is responsible for AIDS is remarkably precise: on average, one error is made in replicating 10,000 base pairs. Because the virus is 10,000 base pairs in size, an error rate this large causes (on average, again) every viral particle to differ from its parent. Remarkably variable in one sense; remarkably stable (one error in 10,000 bases!) in another.

Among sexually reproducing higher organisms, progeny arise from fertilized eggs—that is, from cells that contain DNA contributed by two parents. These DNAs may differ in details of their composition, but, despite these differences (and the resultant huge number of gene combinations that meiotic crossing-over can produce), conservation among successive generations is the rule.

Molecular biologists now possess the means by which genes from one organism can be inserted into the genome of another (horizontal transmission) as a routine exercise. The individual who carries the foreign DNA is called *transgenic*. Thus, a mouse that possesses the growth hormone of a rat (Figure 10-3) is a transgenic mouse. Instances are known in which a naturally occurring horizontal transmission of genetic material is suspected, perhaps following a viral infection; thus, a monkey virus (SSV, simian sarcoma virus) and a cat virus (Pl-Fe Sv, feline sarcoma virus) share nearly identical oncogene portions.

The example of transgenic studies cited below is one carried out by Rollin C. Richmond and his colleagues (Brady et al., 1990; Brady

Figure 10-3. A transgenic mouse carrying an active rat growth-hormone gene (rear) and a normal mouse. The transgenic animal is nearly twice normal size.

and Richmond, 1990). It involves the transformation of *Drosophila melanogaster* with an Esterase gene from *D. pseudoobscura,* a rather distant relative of *D. melanogaster.* The techniques employed in this study will be *sketched* in some detail in order to convey a sense of how the study progressed (and why), but not in the detail needed to transform our readers into laboratory technicians. The procedures that are used to move whole genes from one organism to another will give the curious some insight as to the means (left unexplained earlier) by which 51 base pairs coding for 17 amino acids were excised and replaced in the study of β-lactamase in *E. coli* (Tables 10-1 and 10-2).

Restriction enzymes

Under certain circumstances, bacteria can take up DNA molecules from their aqueous environment and incorporate them into their chromosomes. The bacterium responsible for gonorrhea (*Neisseria gonorrhoeae*) is notoriously adept in this respect. Bacteria are also the recipients of the DNA that is injected into them by attached bacteriophage. Defense against viral infection may occur through the mutational change of surface proteins; these are no longer "recognized" by the tail-filament proteins of the bacteriophage. Internal defense also exists in the form of restriction enzymes. These enzymes, upon attaching to and traveling the length of DNA, recognize it as either "belonging" (usually by the pattern of methylated cytosine bases) or as "foreign." In the latter case, the enzyme cleaves the invading molecule, usually at short palindromic sites (Table 10-3). The different restriction enzymes are identified in Table 10-3 by names

Table 10-3. Recognition, cleavages, and length of single-strand overlap for various restriction enzymes. The names of the various enzymes indicate the source organism: *Escherichia coli* (*Eco*), *Haemophilus influenza* (*Hin*d), *Serratia marcescens* (*Sma*), and other microorganisms.

Enzyme	Restriction site	Length of overlap (base pairs)
*Bam*HI	GGATCC CCTAGG	4
*Eco*RI	CTTAAG GAATTC	4
*Eco*RII	CGGACCG GCCTGGC	5
*Hin*dII	CAPuPyTG GTPyPuAC	blunt end
*Hin*dIII	TTCGAA AAGCTT	4
*Hae*III	CCGG GGCC	blunt end
*Hpa*II	GGCC CCGG	2
*Pst*I	GACGTC CTGCAG	4
*Sal*I	GTCGAC CAGCTG	4
*Sma*I	GGGCCC CCCGGG	blunt end
*Xba*I	TCTAGA AGATCT	4
*Bam*I	CCTAGG GGATCC	4
*Bgl*II	TCTAGA AGATCT	4

that recall the organism from which they were originally obtained (one now buys them by the milligram from chemical supply houses). The recognized site of cleavage (restriction site) may be as short as four (palindromic) base pairs, commonly as long as six or seven, but also (not shown on the table) as long as twenty. The two strands of DNA may be cut at the same position (blunt ends result) or at equal distances from the palindrome's point of symmetry ("staggered" cuts), but always within the recognition site. Shown in Table 10-3 are single-stranded "tails" (overlap) that are two, four, and five base pairs long.

These restriction enzymes (especially those of Type II, which make the "staggered" cuts described above) have proved to be exceptionally useful to molecular geneticists. They are the scissors by which DNA is cut, not at random, but at particular sites. In the following sequence of 120 base pairs that code 40 amino acids, there are two sites at which restriction enzyme *Hpa*II will cleave the DNA molecule, thus creating an intervening sequence of 51 base pairs with two unpaired guanines protruding at one end and a corresponding two unpaired cytosines at the other:

GAA GCC ATG TAC ACC TTT CGA GTT GTG GCT TTG CAA GTT CCA GGG CCA GTA GAA AAC CTG
  E   A   M   Y   T   F   R   V   V   A   L   Q   V   P   G   P   V   E   N   L

AAC GGT CCA GTC CAA GGT TAC AGA TTG TTC CTG GAA GGC CTG AAA AAA TTC ACC GAA TAT
  N   G   P   V   Q   G   Y   R   L   F   L   E   G   L   K   K   F   T   E   Y

Although the pairing attraction of two complementary Cs and Gs is not enormous, in the presence of a ligase (an enzyme that repairs gaps in a DNA molecule), circles of DNA 53 base pairs in circumference could be generated from the excised segment. Each molecule of DNA in the initial solution would, having been exposed to *Hpa*II, generate this small, precisely defined linear segment (or, when ligased, small ring). In an electrophoretic gel, these identical segments would migrate at precisely the same speed and would form a precise band that could be excised and subjected to detailed analyses.

Restriction enzymes that recognize six or more palindromically arranged base pairs have, by chance alone, fewer sites suitable for cleaving. Figure 10-4 illustrates the distribution of restriction sites for six different restriction enzymes that occur in an 11.1 kilobase segment of DNA from *Drosophila pseudoobscura*. This DNA includes the structural genes that code for three Esterases (Esterase-5A, -5B, and -5C). The lower shaded area is 2544 base pairs long and includes the structural gene *Est*-5B (1635-base-pair exon; 55-base-pair intron). Note those numbers—2544, 1635, and 55! They are not approximations. They are precise because the sites on the longer DNA filament that are vulnerable to attack by restriction enzymes have precise locations, and the action of these enzymes themselves is precise.

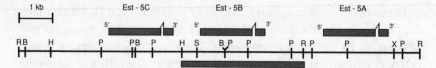

Figure 10-4. Restriction map of the *Est-5* locus of *Drosophila pseudoobscura* and its surrounding DNA. This "locus" contains the structural genes for three related enzymes: EST-5A, EST-5B, and EST-5C (upper case letters represent enzymes; upper and lower case combinations—*Est*—represent genes). The shaded bar below the map represents a segment of the Est-5 locus, 2544 base pairs in length, that contains the *Est*-5B structural gene and some extra DNA at either end; this segment was introduced into *D. melanogaster* by means of P-element transformation (see Figure 10-5). Restriction sites: B, *Bam*HI; H, *Hin*dIII; P, *Pst*I; R, *Eco*RI; S, *Sal*I; X, *Xba*I (see Table 10-3). Inverted *V*s indicate portions of transcribed RNA (introns) that are excised before the remaining RNA (composed of exons) becomes functional mRNA. (After Brady and Richmond, 1990.)

Retaining desired DNA segments

Procedures that are used by molecular geneticists become increasingly sophisticated. Those that are especially valuable are referred to by the names of their inventors or by other, shorthand terms. Still others enter the world of commerce and are known by the company that markets the needed materials and its catalogue number. As a consequence, the account provided here is a generic account, a representation of the well-known "Brand X."

In order to preserve a desired segment of DNA from a fruit fly and, when needed, to have it in respectable quantity, one draws upon *Eschericia coli*, that other favorite organism of geneticists. Plasmids are able to enter individual coliform bacteria, they are able to provide resistance to antibiotics (ampicillin or tetracycline resistance are characteristics of a widely used plasmid), and they can be "persuaded" to carry extra DNA.

If a mixture of plasmids and foreign (e.g., *Drosophila*) DNA is exposed to a restriction enzyme for which there is only a single vulnerable site on the circular plasmid DNA, the mixture will come to possess opened circles (cleaved plasmids) and linear fragments of the foreign DNA. Because of the staggered cuts made by the restriction enzyme, the ends of both plasmids and foreign fragments will be complementary:

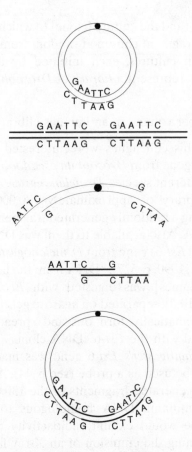

Removal of the restriction enzyme and exposure of the mixture to a ligase will generate many kinds of molecules, including reconstituted plasmids, rejoined fly DNA, and (as illustrated above) plasmids into which fly DNA has been inserted.

This hodgepodge of DNA is now mixed with a culture of *E. coli* (whose cell walls have been temporarily rendered permeable to large molecules), all of which are sensitive, let's say, to ampicillin. After an appropriate exposure, these bacteria are spread on ampicillin-containing media. Only a few colonies of bacteria will survive exposure to the antibiotic; these represent bacteria that took up a plasmid. These few colonies are then tested to determine the nature of the newly introduced plasmid. Plasmids that are larger than they should be (and are not merely two plasmids, joined) contain fly DNA. Each infected bacterium (and the culture [clone] to which it gives rise) contains a restriction fragment chosen at random from the fly's genome. Thousands of such cultures, each initiated by a single plasmid-infected bacterium, represent a *library* of *Drosophila* DNA.

Finding the proper volume in an immense library

Richmond and his colleagues were interested in learning how, when, and where a gene from *Drosophila pseudoobscura* would function in flies of a different species, *D. melanogaster*. Available to these workers was a library of approximately 130,000 volumes—each volume representing a randomly generated fragment of *D. pseudoobscura* genomic DNA. Also available to them was DNA known to represent an Esterase (*Est*-6) gene from *D. melanogaster*. In the authors' words (Brady et al., 1990, p. 527): "DNA from the lambda clones [i.e., the individual volumes] was digested with *Eco*RI [a restriction enzyme, see Table 10-3], separated on agarose gels [the pages of each volume were systematically torn out and spread out for study], blotted, and probed with the *Est*-6 cDNA clone." The cloned DNA (cDNA) of *D. melanogaster's Est*-6 gene was made radioactive in order that it could be used as a probe (see p. 44). As a consequence, only those *Eco*RI-generated fragments of the 130,000 *D. pseudoobscura* volumes containing DNA homologous to that of the *D. melanogaster* probe would exhibit radioactivity (and reveal their presence by darkening the emulsion of an X-ray film). This, then, is the source of the 2544-base-pair segment of DNA diagrammed in Figure 10-4.

Inserting a *D. pseudoobscura* gene into *D. melanogaster*

In the previous chapter, P elements were described briefly (Figure 9-12) as one of the transposable elements found within a higher organism, *D. melanogaster*. These elements are found in some geographic (and laboratory) strains of flies, which are, for that reason, designated *P* strains. They are not found in other (*M*) strains. Hybrid dysgenesis results from crossing *P*-strain males with *M*-strain females (see Figure 9-13), but not in the reciprocal cross.

The normal P-element DNA contains, in addition to an "insertion" sequence of base pairs, a structural gene that codes for a transposase,

an enzyme whose action can cause repeated insertions of its surrounding element at numerous (30–50) sites within the fly's genome
(see p. 209). The insertion sequences are valuable for molecular
geneticists studying the development of or the control of gene action
within *Drosophila:* they offer an opportunity to insert foreign DNA
into the chromosomal DNA of a fly. Consequently, P elements have
been cloned into bacterial plasmids and are maintained in bacterial
cultures. Defective P elements that lack transposase are especially
valuable for research because, once inserted (together with foreign
DNA) into the chromosome of an *M*-strain fly, the inserted element
remains as a single "gene" at the site where it was initially inserted.
It does not move about and increase in number as it would if it also
manufactured transposase.

Figure 10-5 illustrates the procedure by which foreign DNA is
inserted into *D. melanogaster*. The foreign DNA is inserted into a
defective P element; thus, this DNA comes to be flanked by the P
element's insertion sequences. (For Richmond and his colleagues, the
foreign DNA was the lower shaded, 2544-base-pair-long bar containing the gene *Est*-5B of *D. pseudoobscura;* Figure 10-4). This DNA
is injected into very young *Drosophila* embryos by means of a
micropipette. By chance, the injected DNA will enter one or another
cell of the developing embryo and, by virtue of the insertion
sequences and the cell's normal DNA repair machinery, become
incorporated into the chromosomal DNA of that cell. If the cell
happens to be an early germ cell, the foreign DNA—that is, the *Est*-
5B gene in Richmond's case—will be present in the fly's gametes (see
Table 10-4).

Fortunately, the frequency of transgenic adult flies is high enough
($\approx$1%–2%) that screening for them among nontransformed progeny
is not impossible. Indeed, the task is made easier by the presence of
a wild-type allele (ry^+, *rosy*) carried by the transforming plasmid.
When this plasmid is injected into homozygous *ry/ry* embryos, the
eyes of the transformed adults have a deep red rather than reddish
brown color. Having (1) identified these transformed flies and (2)
established them as breeding stocks of flies, one tests (in Richmond's
study) for the gene product of *Est*-5B (of *D. pseudoobscura* origin)
among the treated *D. melanogaster*. The various Esterases are identified by their differing electrophoretic mobilities. The upper portion
of Figure 10-6 reveals the differences between untransformed *D.*

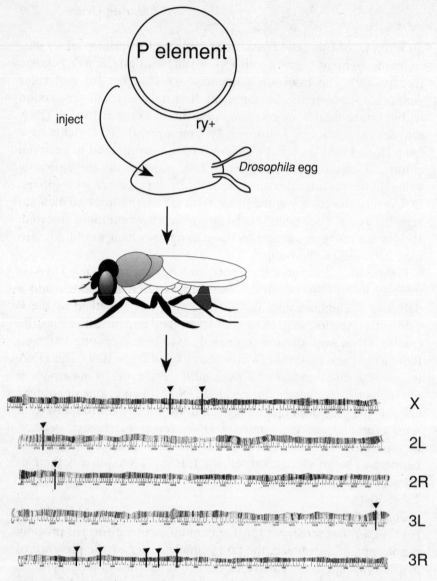

Figure 10-5. P-element transformation in *Drosophila*. P elements containing fragments of *D. melanogaster* DNA carrying the wild-type rosy gene (*ry⁺*) are injected into unhatched *D. melanogaster* embryos (*ry/ry*, rosy eye color). Fertile adults are mated with *ry/ry* flies; those producing wild-type progeny are used in establishing stock cultures of flies in which the *ry⁺* gene has been incorporated into one of the flies' chromosomes. Its precise location can be determined by probing a salivary gland cell giant chromosome with an appropriate, radioactively labeled fragment of DNA. The arrows and heavy bands (revealing the presence of radioactive probe) indicate where P elements have been known to insert into the *Drosophila* genome.

Table 10-4. Summary of data obtained from six experiments in which mutant (*rosy* eyes) *Drosophila melanogaster* embryos were injected with defective P elements carrying the wild-type allele of the *rosy* gene. Fertile adults that produced wild-type (redrather than rosy-eyed) progeny must have acquired an infective P element in their germ cells. In subsequent generations, the wild-type allele is inherited in these lines as a Mendelian gene but not necessarily at the same locus (or even the *rosy* locus) in every transformed line. (After Rubin and Spradling, 1982.)

Experiment no.	No. of embryos injected	No. of fertile adults	No. of adults giving ry^+ (red-eyed) progeny
1	147	9	2
2	114	5	1
3	215	19	11
4	121	17	5
5	87	4	2
6	427	28	0

melanogaster, transformed *D. melanogaster* carrying the *Est*-5B gene of *D. pseudoobscura,* and wild-type *D. pseudoobscura.* The outermost lanes reveal that flies of the two species have Esterases of different electrophoretic mobilities; the transformed (= transgenic) *D. melanogaster* have both enzymes.

Obviously, the gene from *D. pseudoobscura* functions when transferred into *D. melanogaster;* that has been revealed by the enzymatic tests of squashed flies shown in Figure 10-6. How about the pattern of expression among various tissues, however? In *D. melanogaster,* most of the gene product of the *Est*-6 locus is found in the male fly's ejaculatory duct; the corresponding gene from *D. pseudoobscura* makes an Esterase that is found in the compound eyes (40%) and in the body fluid hemolymph (≈60%). Figure 10-7 reveals that isolated reproductive systems of *D. melanogaster* males exhibit Esterase-6 activity; neither transgenic *D. melanogaster* nor *D. pseudoobscura* males' reproductive systems exhibit Esterase-5 activity. In the case of hemolymph, Esterase-5 exhibits activity in the hemolymph of transgenic *D. melanogaster* as it does *D. pseudoobscura.* The *Est*-6 allele produces its enzyme in the hemolymph of both wild-type and transgenic *D. melanogaster.*

One gathers from Figure 10-7 that the *D. pseudoobscura* gene, even though it now rests in a *D. melanogaster* fly, functions in only those

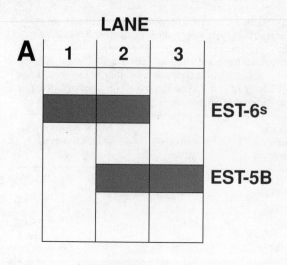

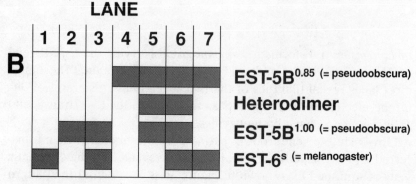

Figure 10-6. The presence of two Esterases (EST-5B and EST-6^s) in polyacrylamide gels following electrophoresis. The gene *Est*-5B, is normally present in *Drosophila pseudoobscura; Est*-6^s is normally present in *D. melanogaster*. The similarity of the base composition of the DNA coding for these two enzymes suggests that they are homologous. (A) Lane 1, nontransformed *D. melanogaster;* lane 2, transformed *D. melanogaster* carrying the *Est*-5B$^{1.00}$ gene of *D. pseudoobscura;* lane 3, *D. pseudoobscura* expressing the Esterase, EST-5B. (B) Lane 1, untransformed *D. melanogaster;* lane 2, *D. pseudoobscura* expressing a fast-moving variant of its normal gene, *Est*-5B$^{1.00}$; lane 3, transformed *D. melanogaster* expressing Esterases EST-6^s and EST-5B$^{1.00}$; lane 4, *D. pseudoobscura* expressing a slow-moving variant of the enzyme EST-5B$^{0.85}$; lane 5, a crushed mixture of *D. pseudoobscura* (*Est*-5B$^{1.00}$) and *D. pseudoobscura* (*Est*-5B$^{0.85}$) (note the formation of *heterodimers*—that is, of Esterases that consist of polypeptide chains from both crushed flies); lane 6, a crushed mixture of transformed *D. melanogaster* (*Est*-6^s/*Est*-5B$^{0.85}$) and *D. pseudoobscura* (*Est*-5B$^{0.85}$) (note the presence of EST-5B/EST-6 "interspecific" heterodimers that have been formed within the transformed *D. melanogaster*); lane 7, *D. melanogaster* (*Est*-6^s) and *D. pseudoobscura* (*Est*-5B$^{0.85}$) (heterodimers do not form between the polypeptides when flies of the two species are crushed together).

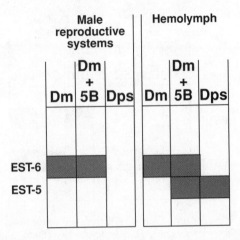

Figure 10-7. The presence or absence of Esterase activity in (left) the male reproductive systems and (right) the hemolymph of nontransformed *Drosophila melanogaster* (Dm), transformed *D. melanogaster* (Dm + 5B), and *D. pseudoobscura* (Dps). The Esterase EST-5B does not normally appear in the male ejaculatory ducts of *D. pseudoobscura*, nor does it in those of transformed *D. melanogaster* males. In contrast, EST-6 does normally appear in the ducts of *D. melanogaster* males. EST-5B is found in large quantities in the hemolymph of *D. pseudoobscura*; it also appears in the hemolymph of transformed *D. melanogaster*, together with the enzyme EST-6, which is normally found in the hemolymph of that species.

tissues in which it functions when in the proper milieu. This conclusion is extended by the tests illustrated in Figure 10-8 and listed in Table 10-5. Esterase-6 does not normally occur in the eyes of *D. melanogaster,* although it is found in the head tissue other than eyes. All electrophoretic gels involving *D. melanogaster* (transgenic or not) exhibit this normal pattern for Esterase-6 expression. Esterase-5B, on the other hand, normally occurs in both the eyes and in the remaining head tissue of *D. pseudoobscura;* this same pattern is found even when the gene is transferred to another species (Dm+5B). A cursory examination of the many tissues listed in Table 10-5 reveals that the activity (+ or –) of Esterase-5 in transgenic *D. melanogaster* closely matches that for the same enzyme in its natural "home"—*D. pseudoobscura.*

The conclusion reached by Richmond and his colleagues is that the activity of the *Est*-5B gene resides with the gene itself and its accompanying nearby DNA. Had signals come to the locus from elsewhere (*trans*-acting), the *Est*-5B gene transferred into *D. melanogaster* might

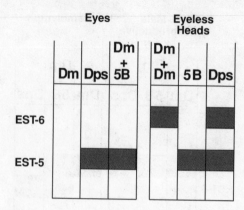

Figure 10-8. The presence or absence of Esterases in the eyes or eyeless heads of *Drosophila melanogaster* (Dm), *D. pseudoobscura* (Dps), and transformed *D. melanogaster* (Dm + 5B). EST-6 does not normally occur in the eyes of *D. melanogaster,* nor does it do so in transformed flies. EST-5B does occur in the eyes both of *D. pseudoobscura* and of transformed *D. melanogaster*. EST-6, which is found in the heads (not eyes) of *D. melanogaster,* is found in the heads of transformed *D. melanogaster* as well. EST-5B, which normally is found in the head (as well as eyes) of *D. pseudoobscura,* is also found in the eyeless heads of transformed *D. melanogaster*.

have been expected to act in concert with the "resident" Esterase-6. It did not do so in any instance studied. The material chosen to be presented here as a "case study" may, of course, have its idiosyncracies; studies involving other genes and other organisms may yet reveal an inability of a foreign gene to act after transferral to an alien species because it cannot "read" that species' regulatory signal. An inability of the operator at the *lac* locus in *E. coli* to respond normally to the repressor substance is, as we saw in Chapter 9, a frequent cause of the abnormal functioning of that locus.

The polymerase chain reaction

The polymerase chain reaction, otherwise known as PCR, is our final example of modern techniques that have been developed by molecular geneticists. It is an extremely powerful tool in the study of gene action as we shall illustrate shortly. It is a technique, however, that is appropriate for many studies. Two of these, the phylogenetic

Table 10-5. Relative activities of EST-6 and EST-5 in dissected tissues from 12 adult flies or third-instar larvae. (After Brady and Richmond, 1990.)

Tissue	Isozyme	Dm	Dm+5B	Dps
			Esterase activity*	
			Adults	
Total gut	EST-5		+	+
	EST-6	+	+	
Crop	EST-5		+	+
	EST-6	+	+	
Malpighian tubules	EST-5		+	+
	EST-6	+	+	
Ovaries	EST-5		++	++
	EST-6	++	++	
Body wall	EST-5		+	+
	EST-6	+	+	
			Larvae	
Brain	EST-5		—	—
	EST-6	—	—	
Midgut	EST-5		—	—
	EST-6	—	—	
Hindgut	EST-5		—	—
	EST-6	—	—	
Malpighian tubules	EST-5		—	—
	EST-6	—	—	
Salivary glands	EST-5		—	—
	EST-6	—	—	
Fat body	EST-5		—	—
	EST-6	—	—	
Tracheae	EST-5		—	—
	EST-6	—	—	
Hemolymph	EST-5		+++	+++
	EST-6	+++	+++	
Carcass	EST-5		+	+
	EST-6	+	+	

* Activities were measured by analysis of nondenaturing polyacrylamide gels. —, no activity; +, low activity; ++, moderate activity; +++, high activity. Dm, untransformed *Drosophila melanogaster;* Dm+5B, *D. melanogaster* transformed with *Est-5;* Dps, *D. pseudoobscura.*

relationships among organisms and the study of genetic variation in populations, can be mentioned even before describing the laboratory procedures in some detail.

PCR is a procedure that permits a person to take a small piece of

DNA and, by a clever use of heat-resistant DNA polymerase, to double and re-double the number of copies of that piece approximately every five or six minutes. The number of copies in a test tube, that is, grows exponentially just as bacteria do in a bacterial culture— but with a much shorter generation time. Within an hour or two after preparing the initial material, one can obtain many million, virtually identical copies of the initial (single) segment of DNA.

Without referring to individual research papers, one can mention situations in which the PCR procedure can be extremely useful. Small amounts of DNA can be recovered from mammalian hair. Thus, single hairs removed from museum pelts that have been on display (or stored out of sight) for centuries can be used to study the genetic variation present in ancient populations of fur-bearing beasts. Among the fur pieces that are left unharmed by the removal of single hairs are the furs that decorated royal robes: cuffs, neck pieces, and other trimmings.

Fossil organisms of many sorts are preserved in manners that do not completely destroy DNA. Thus, DNA can be recovered from mammoths that have been preserved in the frozen arctic tundra, amplified by the polymerase chain reaction, and then compared with the DNA of modern African and Asian elephants. S. J. O'Brien, a specialist in the molecular phylogeny within the cat family, has recently studied samples of DNA obtained from the fossil saber-toothed tigers trapped in the La Brea tar pits of Los Angeles; these fragments of DNA, increased many-fold by PCR, have allowed O'Brien and his colleagues to place the great saber-toothed cats into the phylogeny of modern-day cats (big, little, and medium-sized) with extreme accuracy (Janczewski et al., 1992). They are panthers! Insects that have been trapped in amber for 100 million years have yielded DNA that is amenable for PCR amplification. (DNA recovered from dinosaur blood cells in the stomachs of amber-embedded mosquitoes is still a matter for writers of fiction.)

A final study can be mentioned here. Population genetics is a science that attempts to detect, measure, and account for the presence of genetic variation in populations, natural populations for the most part. Changes in the nature and amount of this variation through time fall within the scope of population genetics; for this reason, population geneticists are deeply involved in the study of evolution. The duration of present-day population studies is extremely short in evolutionary terms. A tremendously detailed study of the native snails of Pacific islands that was begun about 1900 has been ended by the virtual extinction of those snails by an imported carnivorous one. Observations on *Drosophila pseudoobscura* that were begun by Th. Dobzhansky and A. H. Sturtevant 60 years ago are being continued and extended by Dobzhansky's students (many of whom are now in their late fifties); although 60 years is a long time, it is but an instant with respect to evolutionary change. Real estate developers and atmospheric pollution threaten to destroy the sites at which these flies were originally studied.

Consider now what kind of study PCR may make possible. D. F. Owen (1966) reported that fossil snails (*Limicolaria martensiana*) of Africa are sufficiently well preserved that the variation in the color types of fossil shells can be determined; these snails—8000 to 10,000 years old—possessed pattern polymorphisms just as many present-day snail species (e.g., *Cepaea hostensis* and *C. nemoralis*) do. If small amounts of DNA have been preserved in and can be recovered from these fossil snails, PCR allows (at least in theory) the student of snail populations to extend the analysis of the genetic variation of populations back into past centuries and millenia.

Experimental procedures

The procedures that lead to the polymerase chain reaction in replicating DNA are listed in Figure 10-9. The steps illustrated in the figure are not difficult to follow. What is not explained there is this: Where do the primer sequences (Figure 10-9B) come from? These sequences are made by the investigator, using his or her intuition and background knowledge. Using the fossil snail study as an example, one might turn to a library of cloned flour beetle (*Tribolium*) DNA and, with radioactively labeled DNA, probe for that portion that specifies a certain enzyme or that specifies ribosomal RNA. Whatever segment of DNA is "withdrawn" from the library, its entire sequence is determined. On the basis of that sequence, the investigator then synthesizes the primers in the laboratory; as a rule, primers are 20–30 base pairs long, and they are chosen so that they are separated by several hundred intervening base pairs.

A. Original DNA

▶TACGGAGCTGGCGGAATTATTA▶
◀ATGCCTCGACCGCCTTAATAAT◀

B. Primer molecules

▶GAGC▶   ▶AATT▶
◀CTCG◀   and   ◀TTAA◀

C. Denature at 94° (90 sec.): Both original and primer DNA melt to form single stands

D. Reanneal at 55° (2 min.): Will generate, among others

▶TACGGAGCTGGCGGAATTATTA▶
                  ◀TTAA◀

and

◀ATGCCTCGACCGCCTTAATAAT◀
▶GAGC▶

E. Extend DNA by polymerase at 72° (3 min.)

▶TACGGAGCTGGCGGAATTATTA▶
◀ATGCCTCGACCGCCTTAA◀

and

◀ATGCCTCGACCGCCTTAATAAT◀
▶GAGCTGGCGGAATTATTA▶

F. Upon remelting and repeating the cycle numerous times, the most common DNA molecules (because of exponential increase) becomes:

▶CTCGACCGCCTTAA▶
◀GAGCTGGCGGAATT◀

NOTE: The common DNA segment includes the primer sequences and the intervening base pairs, of which there may be a thousand or more.

Figure 10-9. A more or less narrative presentation of the polymerase chain reaction (PCR). The example used has been grossly simplified. The primer sequences are usually 20–30 base pairs long. The intervening base pairs may number in the hundreds (numbers greater than 1000 may pose technical difficulties). The polymerase introduces errors (10^{-4}), but these can be allowed for in most studies. (Not so, however, in reconstructing phylogenies for closely related species, in which conclusions are based on slight differences in base-pair composition!) The constant shifting of temperatures from 94°C to 55°C to 72°C, repeated in every cycle, quickly destroys many polymerases (being proteins, they become denatured). This problem is overcome by using polymerases obtained from thermophilic bacteria; some of these polymerases can function at temperatures exceeding 100°C. The arrows indicate the 5′ to 3′ direction of the DNA strand; this is the direction in which the polymerase extends the strand from the primer attachment site.

Having gotten the primers as a preliminary exercise, one can now begin with DNA obtained from any source—the fossil snails, in the present case. The primers (in great excess) are added to a mixture that contains the DNA to be studied, the DNA polymerase, deoxyribonucleoside triphosphates (these provide the bases that the polymerase will add to the DNA-primer complex), and various buffer salts.

The DNA molecules are denatured (melted) at 94°C (a step that requires about 90 seconds). The temperature of the mixture is then reduced to 55°C. Within two minutes, complementary DNA strands have reannealed with one another. Many of these create reconstituted primer molecules, but these are of no interest. Others (as illustrated in Figure 10-10D) will consist of a single strand of snail DNA paired at some point with a much shorter primer molecule. The excess concentration of primer favors this pairing over the reannealing of the much longer snail DNA molecules.

When the temperature of the mixture is raised once more (but only to 72°C), the DNA polymerase molecules become active and, passing down (5' to 3') the long single-stranded molecule that protrudes beyond the primer, add complementary bases to the still-unpaired single strand. To add the complementary base to a strand 2000–3000 bases long requires about 3 minutes.

At the end of that time (about 6 minutes in all), one can reheat the mixture to 94°, thus melting the DNA molecules again. Once the mixture is cooled to 55°, the separated strands reanneal—many with primers attached to much longer strands as before. Raising the temperature to 72° reactivates the polymerase again, thus leading to the synthesis of complementary strands of DNA. The original, long molecules of snail DNA are not destroyed, so they continue to be copied. The first-generation copies that lose one or the other of the raw ends that extend beyond the primer are also copied in later cycles. However, these DNA molecules increase in number during successive cycles in an arithmetic progression: 2, 4, 6, 8, 10, . . . The DNA molecules that terminate at the distal ends of the primers increase in geometric progression: 2, 4, 8, 16, 32, . . . Thus, after 30 cycles (approximately 3 hours) the mixture should contain fewer than 300 unwanted molecules and 2.68×10^8 of the precisely made ones. These can be seen (and isolated) as a clear sharp band by means of gel electrophoresis.

Applications of PCR for the study of gene action

The polymerase chain reaction represents one of the most useful (and, therefore, exciting) technical advances in the study of gene action. As outlined by Donald M. Coen (1991), PCR allows the amplification of DNA from either genomic or cloned DNA. It permits the amplification of mutagenized DNA molecules so that those with single changes can be obtained in quantity. To reinsert these altered molecules into the donor organism is not an impossible task, as we have seen. With proper probes, one can test for viral DNA that may or may not be incorporated into genomic DNA. One can isolate ribosomes, expose them to antibodies that recognize a particular protein, isolate the ribosomes that are producing that protein, and remove the mRNA that is being translated into that particular protein. That mRNA, by means of reverse transcriptase, can be used to generate the corresponding (and more stable) DNA in tiny amounts. The polymerase chain reaction, however, can be used to convert those tiny amounts of DNA into relatively huge quantities. (This entire procedure is referred to by the abbreviation RT-PCR.) The DNA obtained in this way represents the gene that specifies the chosen protein. Inserted into *E. coli,* this gene can be put to work, synthesizing its protein product in commercial quantities. The applications of PCR seem limitless; more appear daily.

11 Epilogue

And here we are! To say that we have reached the shortest string of our metaphorical harp (Figure 1-1) would be not only presumptuous but wrong. In their preface to the fourth edition of *Molecular Biology of the Gene* (Watson et al., 1987, p. v), J. D. Watson and his colleagues write:

> It is only in this fourth edition that we see the extraordinary fruits of the recombinant DNA revolution. Hardly any contemporary experiment on gene structure or function is done today without recourse to ever more powerful methods for cloning and sequencing genes. As a result, we are barraged daily by arresting new facts of such importance that we seldom can relax long enough to take comfort in the accomplishments of the immediate past. The science described in this edition is by any measure an extraordinary example of human achievement.

Under circumstances in which, as these authors say, "We are barraged daily by arresting new facts of such importance that we seldom can relax," one senses the speed with which the harp is being extended to accommodate ever shorter strings—strings that grow in importance, however, with ever greater rapidity. In the previous chapter, PCR was introduced not in conjunction with the role it plays in the study of gene action but, rather, with reference to its application to evolutionary problems—phylogeny and genetic variation within populations. Such problems must be among the longest strings on our harp. At least Mendel discussed the reapportionment of genetic variation within a population of self-fertilizing plants.

No textbook designed for the merely intellectually curious, for the post–general biology nonmajor, can hope to be technically up to date. The quotation at the end of Chapter 10 (Coen, 1991) is from a publication titled *Current Protocols in Molecular Biology. Current Protocols* is published in loose-leaf form and contains many chapters and supplements. That is the format needed to keep professionals *au courant*. Those who watch from the sidelines must be satisfied with the program that identifies the players and outlines the general rules of the game.

In making the concluding remarks for this text, we must admit that the assessment given by Watson and his colleagues is correct: molecular genetics is, indeed, an extraordinary example of human achievement. During the century following the rediscovery of Mendel's paper in 1900, geneticists relentlessly pursued the gene itself, until they demonstrated that it resided in (or consisted of) DNA—a substance whose structure was once thought to be too simple for any purpose except for chromosomal scaffolding (see Wallace, 1992). During the early decades of the twentieth century, some geneticists more than others were concerned with the manner in which genes perform their functions (these persons were content for the most part to assume—erroneously—that the gene itself was an autocatalytic enzyme). It is this branch of genetics that we have traced in the present text. And it is this branch that has exploded into prominence following the discovery of the true nature of the gene, for it is that structure that has led to the splendid genetic manipulations that occur daily in thousands of laboratories, worldwide.

In their prefatory remarks, Watson and his colleagues (Watson et al., 1987, p. v) refer to the early concern of many scientists—including geneticists—"that recombinant DNA procedures might generate dangerous and pathogenic new organisms." This matter was debated in numerous letters to editors, editorials, and scientific meetings during the early 1970s. The culminating conference was held at the Asilomar Conference Center, California, during February 1975. The recommendations that emerged from that conference (DNA should be cloned only into genetically defective ["safe"] organisms; special laboratory precautions should be observed) formed the bases for guidelines subsequently issued by the National Institutes of Health.

The logic that underlies the effectiveness of the immune systems of higher vertebrates is not unrelated to the question of "safety" that is raised by recombining the DNA of diverse organisms. Each person's health depends upon the random (or seemingly haphazard) reorganization at the molecular level of the gene(s) that specify antibody structure. The logic is (and it works) that given an enormous variety of antibody proteins, at least one will almost surely interact with any foreign organism or substance that invades one's body. Once that initial interaction occurs, other mechanisms exist by which the interacting antibody is manufactured in quantity. As we stated in Chapter 9, rare events can become near certainties if their opportunities become enormous.

This same reasoning applies to recombinant DNA research. DNA is no ordinary chemical: It directs its own replication. It can change. And the changed form directs its replication as diligently as the original molecule did. With considerable justification, Muller (1960) referred to these abilities as defining "life"; the characteristics more generally cited in distinguishing living organisms from nonliving objects are, according to Muller, characteristics that have been acquired through natural selection because they facilitate and enhance the successful reproduction of the responsible DNA. At any rate, however small the risk that any new recombinant DNA might pose a threat to the preexisting natural world (including human beings), an "accident" is virtually certain to happen as the absolute number of recombined DNA molecules of different sorts are generated. Our own ability to fight off infectious diseases assures us that this is so.

An entire constellation of ancillary questions other than the possibility of an ecological disaster arises in conjunction with studies made by molecular biologists, whether such studies are carried out for the public's welfare or for personal gain. There is no compelling need for these questions to be phrased anew here. David Suzuki and his colleagues (Suzuki et al., 1986, p. 328) have expressed matters in a thoughtful manner in the three paragraphs quoted here:

> Nonetheless, the turmoil about recombinant DNA did raise important social issues. For example, what is the social responsibility of scientists who are developing powerful new technologies? Should the people who are doing the experiments be the ones who set the guidelines? At what point should the public have an input? Who should be legally liable for any accidental damage that results from scientific research?

Should limits be placed on the freedom of scientists to design and conduct research projects? Should a scientist attempt to foresee possible adverse effects from future use of discoveries and refuse to advance knowledge in certain directions that might have unfortunate applications?

Other kinds of questions are raised by the controversy over recombinant DNA. Can we predict with confidence the properties of an organism modified by inserting DNA from a totally unrelated source? Can there be deleterious effects that will not be detected until large populations have been exposed for years (as was the case with oral contraceptives)? In the long run, will increasing sophistication in DNA manipulation inevitably lead to genetic manipulation of human beings? If so, who will decide the conditions?

Although much of the worry about dangers of recombinant DNA research has been laid to rest, the issue has served to raise far more profound questions about the relationship between science and society. These questions have not been answered satisfactorily, and they are likely to persist and become even more important in the coming years.

Final points concerning modern research in molecular genetics to be discussed in terminating this chapter can be subsumed under the umbrella term *resource allocation*. Modern biological research is expensive! Following World War II, during the late 1940s and early 1950s, genetic research in the United States (including the salaries of six technicians and the principal investigator) cost between $25,000 and $35,000 annually. As late as the mid-1980s, the chairman of a visiting "site-visit" committee could chastise a university departmental chairman for allowing his faculty to concentrate entirely on individual research grants of approximately $35,000 each (not with seven salaries included, however!) to the neglect of a cooperative department-wide training grant that might involve $2 million or $3 million over a period of five years. Today, a modest laboratory in which the principal investigator spends much of his or her time, together with a laboratory technician, requires $150,000 or more annually. A young molecular biologist seeking a new position in an "unprepared" university department might request $500,000 to $750,000 in "setup" funds. The Human Genome Project is estimated to cost $3 billion; yeast, nematode, and *Arabidopsis* (a small plant) genome projects cost less—but are still expensive.

Clearly, when individual states must place municipal bond issues involving mere hundreds of millions of dollars before the voters for approval, a research program in molecular genetics costing $3 billion ($3000 millions) must be viewed as a major intrusion into a nation's resource base. Large as the medical research budget of the National Institutes of Health may be, an expenditure of $3 billion cannot avoid depleting funds available for smaller, individual research efforts.

Still other claims are being billed against the nation's research funds. A multitude of ills—the destruction of the earth's ozone layer, the devastation being wrought in the world's tropical rain forests, oceanic oil spills that eventually cover thousands of miles of marine estuaries and beaches, the pollution or exhaustion of aquifers for irrigation and domestic or commercial use, and the extinction of untold numbers of plant and animal species—have made all of us aware of environmental problems. Many ecologists and their colleagues who are engaged in organismic biology have endorsed the creation of a National Institute of the Environment, an institute whose budget would be comparable to that of the National Institutes of Health.

In persuading governmental officials at all levels that molecular biology, including the Human Genome Project, is worthy of considerable support, advocates of such research stress the importance of curing or preventing a multitude of human ills—cancer, genetic diseases, and immunologic disorders. Again, the specter of resource allocation arises, but at the individual level rather than at the national or community level. To simplify this, the concluding, discussion, we might agree that if shots comparable to those that are given routinely to growing children can be developed for the prevention or cure of genetic diseases, molecular biology will have repaid its cost to society. On the other hand, if the treatments that emerge from molecular research are expensive and available only to the very well-to-do, one might question the basis on which such research is funded at its present, rather lucrative level.

Let's imagine the case in which $1 million is available for the treatment of a certain disorder. This money may be spent entirely on one person, or it may be spent on 1 million persons, each of whom would receive a one-dollar treatment. Underlying Figure 11-1 are these specified situations: First, the expenditure of more than $100,000 on an individual does not materially improve the efficiency of the treatment; 10 persons can be treated at this level of care. Second, the expenditure of less than $100 on a patient is essentially the same as

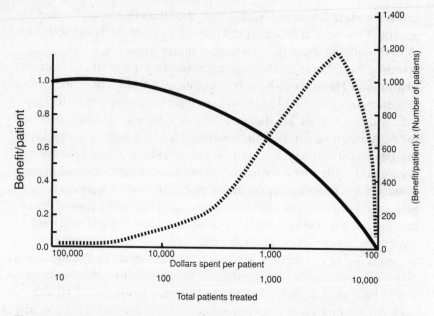

Figure 11-1. The benefits that accrue to both the individual (solid line) and to "society" (dashed line) as limited medical resources ($1 million) are allocated among individuals (from $100,000 to each of 10 individuals to $100 to each of 10,000). It is assumed that no additional benefit to the patient is gained by expenditures exceeding $100,000 and that expenditures less than $100 per patient are virtually useless. The dashed line implies that society benefits most when *individuals* benefit at a level only one-quarter or one-third the maximum possible.

providing no care whatsoever; 10,000 persons can be treated at $100 each but essentially to no avail.

Beginning at the $100 level, the benefit to the treated patient increases as shown by the solid line in Figure 11-1, gradually leveling off as the amount approaches $100,000. What, though, is the cumulative benefit calculated by multiplying benefit per person by the number of persons benefited? This product is shown by the dashed line. The cumulative benefit rises sharply (going from right to left) and then falls once more as the added cost of the treatment has less and less beneficial effect on the effectiveness of the treatment. It would seem from the dashed curve in Figure 11-1 that the greatest good to society as a whole is achieved when the individual is benefited only one-quarter or one-third the maximum possible, but where some 3500 persons rather than 10 are treated.

We now leave it to our student (or other) reader, as someone who is about to assume the responsibilities that accompany participation in a democracy, to decide for him- or herself the proper allocation of funds for health care. Do we as a nation want high-tech medicine that is available for only a fortunate few? Do we want to divert our resources from the homeless, the aged, and the unemployed so that virtually every newborn baby has potential access to teams of surgeons, standing ready to correct or replace its defective organs? The matter is only peripherally related to the study of gene action; nonetheless, that study *is* involved because we inhabit a rather small, finite earth possessing limited resources that we all must share.

References

Anderson, Edgar. 1949. *Introgressive Hybridization*. New York: John Wiley.

Anderson, Wyatt. 1973. Genetic divergence in body size among experimental populations of *Drosophila pseudoobscura* kept at different temperatures. Evolution 27:278–284.

Astbury, W. T., and F. O. Bell. 1938. Some recent developments in the X-ray study of proteins and related structures. Cold Spring Harbor Symp. Quant. Biol. 6:109–121.

Avery, O. T., C. M. MacLeod, and M. McCarty. 1944. Studies on the chemical nature of the substance inducing transformation of pneumococcal types. J. Exp. Med. 79:137–157.

Baker, W. K. 1963. Genetic control of pigment differentiation in somatic cells. Am. Zool. 3:57–69.

Baker, W. K. 1968. Position-effect variegation. Adv. Genet. 14:133–169.

Baker, W. K. 1978. A genetic framework for *Drosophila* development. Annu. Rev. Genet. 12:451–470.

Balbiani, E. G. 1881. Sur la structure du noyau des celles salivaires chez les larves de *Chironomus*. Zool. Anz. 4:637–641.

Beadle, G. W. 1945. Biochemical genetics. Chem. Rev. 37:15–96.

Beadle, G. W. 1963. *Genetics and Modern Biology*. Philadelphia: American Philosophical Society.

Beadle, G. W., and Boris Ephrussi. 1936. The differentiation of eye pigments in *Drosophila* as studied by transplantation. Genetics 21:225–247.

Beadle, G. W., and E. L. Tatum. 1941a. Experimental control of development and differentiation. Am. Nat. 75:107–116.

Beadle, G. W., and E. L. Tatum. 1941b. Genetic control of biochemical reactions in *Neurospora*. Proc. Natl. Acad. Sci. USA 27:499–506.

Becker, H. J. 1957. Über Rötgenmosaikflecken und Defektmutationen am Auge von *Drosophila* und die Entwicklungsphysiologie des Auges. Z. Indukt. Abstammungs Vererbungsl. 88:333–373.

Benzer, Seymour. 1973. Genetic dissection of behavior. Sci. Am., December: 24–37.

Berns, M. W., D. E. Rounds, and R. S. Olson. 1969. Effects of laser micro-irradiation on chromosomes. Exp. Cell. Res. 56:292–298.

Brady, J. P., and R. C. Richmond. 1990. Molecular analysis of evolutionary changes in the expression of *Drosophila* esterases. Proc. Natl. Acad. Sci. USA 87:8217–8221.

Brady, J. P., R. C. Richmond, and J. G. Oakeshott. 1990. Cloning of the Esterase-5 locus from *Drosophila pseudoobscura* and comparison with its homologue in *D. melanogaster*. Mol. Biol. Evol. 7:525–546.

Bregliano, J. C., and M. G. Kidwell. 1983. Hybrid dysgenesis determinants. In J. A. Shapiro, ed., *Mobile Genetic Elements*, pp. 363–410. New York: Academic Press.

Bridges, C. B. 1916. Non-disjunction as proof of the chromosome theory of heredity: I and II. Genetics 1:1–52, 107–163.

Bridges, C. B. 1935. Salivary chromosome maps. With a key to the banding of the chromosomes of *Drosophila melanogaster*. J. Hered. 26:60–64.

Bridges, C. B., and T. H. Morgan. 1923. The third-chromosome group of mutant characters of *Drosophila melanogaster*. Carnegie Inst. Wash. Publ. 327. 251 pp.

Britten, R. J., and E. H. Davidson. 1969. Gene regulation for higher cells: a theory. Science 165:349–357.

Catcheside, D. G. 1939. A position effect in *Oenothera*. J. Genet. 38:345–352.

Cech, T. R., and B. L. Bass. 1986. Biological catalysis by RNA. Annu. Rev. Biochem. 55:599–629.

Chen, T. Y. 1929. On the development of imaginal buds in normal and mutant *Drosophila melanogaster*. J. Morphol. 47:135–199.

Coen, Donald M. 1991. Enzymatic amplification of DNA by PCR: standard procedures and optimization. Curr. Protocols Mol. Biol., Suppl. 16 (Unit 15.1):1–7.

Corcoran, M. R. 1976. Gibberellin antagonists and anti-gibberellins. In H. N. Krishnamoorthy, ed., *Gibberellins and Plant Growth*, pp. 289–332. New York: John Wiley and Sons.

Creighton, H. B., and B. McClintock. 1931. A correlation of cytological and genetical crossing-over in *Zea mays*. Proc. Natl. Acad. Sci. USA 17:492–497.

Crick, F. H. C., and P. A. Lawrence. 1975. Compartments and polyclones in insect development. Science 189:340–347.

Crow, J. F., and M. Kimura. 1965. Evolution in sexual and asexual populations. Am. Nat. 99:439–450.

Darwin, Charles. 1876. *The Effects of Cross and Self-Fertilization in the Vegetable Kingdom*. London: Murray.

Davis, B. D. 1948. Isolation of biochemically deficient mutants of bacteria by penicillin. J. Am. Chem. Soc. 70:4267.

Demerec, M. 1931. Behavior of two mutable genes of *Delphinium ajacis*. J. Genet. 24:179–193.

Demerec, M. 1940. Genetic behavior of euchromatic segments inserted into heterochromatin. Genetics 25:618–627.

Demerec, M., and P. E. Hartman. 1959. Complex loci in microorganisms. Annu. Rev. Microbiol. 13:377–406.

Dobzhansky, Th., and Boris Spassky. 1944. Genetics of natural populations. XI. Manifestation of genetic variants in *Drosophila pseudoobscura* in different environments. Genetics 29:270–290.

Dunn, L. C., and D. R. Charles. 1937. Studies on spotting patterns. I. Analysis of quantitative variations in the pied spotting of the house mouse. Genetics 22:14–42.

Edgar, R. S., and R. H. Epstein. 1965. The genetics of a bacterial virus. Sci. Am., February:70–78.

Emerson, R. A. 1921. The genetic relations of plant colors in maize. Cornell Univ. Agric. Exp. St. Mem. 39. 156 pp.

Emmerling, M. H. 1958. An analysis of intragenic and extragenic mutations of the plant color component of the R^r gene complex in *Zea mays*. Cold Spring Harbor Symp. Quant. Biol. 23:393–407.

Ephrussi, Boris. 1952. The interplay of heredity and environment in the synthesis of respiratory enzymes in yeast. Harvey Lect. (46)1950–51:45–67.

Ephrussi, Boris, and Eileen Sutton. 1944. A reconsideration of the mechanism of position effect. Proc. Natl. Acad. Sci. USA 30:183–197.

Ephrussi-Taylor, Harriet. 1951. Genetic aspects of transformations of pneumococci. Cold Spring Harbor Symp. Quant. Biol. 16:445–456.

Fedoroff, Nina. 1984. Transposable genetic elements in maize. Sci. Am., June:84–98.

Fedoroff, Nina. 1991. Maize transposable elements. Perspect. Biol. Med. 35:2–19.

Fincham, J. R. S. 1985. From auxotrophic mutants to DNA sequences. In J. W. Bennett and L. L. Lasure, eds., *Gene Manipulations in Fungi*, pp. 3–34. New York: Academic Press.

Garrod, A. E. 1909. *Inborn Errors of Metabolism*. Oxford: Frowde Hodder and Stoughton, Oxford University Press.

Goldschmidt, Richard B. 1949. Phenocopies. Sci. Am., October:46–49.

Haldane, J. B. S. 1942. *New Paths in Genetics.* New York: Harper.

Hardin, Garrett. 1964. *Population, Evolution, and Birth Control.* San Francisco: W. H. Freeman.

Horowitz, N. H. 1950. Biochemical genetics of *Neurospora*. Adv. Genet. 3:33–71.

Hotta, Yoshiki, and Seymour Benzer. 1972. Mapping of behaviour in *Drosophila* mosaics. Nature 240:527–535.

Howard Hughes Medical Institute. 1992. *From Egg to Adult.* Bethesda, Md.: Howard Hughes Medical Institute.

Ingham, P. W. 1988. The molecular genetics of embryonic pattern formation in *Drosophila*. Nature 335:25–34.

Janczewski, D. N., N. Yuhki, D. A. Gilbert, G. T. Jefferson, and S. J. O'Brien. 1992. Molecular phylogenetic inference from saber-toothed cat fossils of Rancho la Brea. Proc. Natl. Acad. Sci. USA 89:9769–9773.

John, Bernard, and G. L. G. Miklos. 1988. *The Eukaryote Genome in Development and Evolution.* London: Allen and Unwin.

Kästner, Erich. 1969. Mein Onkel Franz. Copenhagen: Atrium Verlag Zuerich.

Kidwell, M. G., and J. B. Novy. 1979. Hybrid dysgenesis in *Drosophila melanogaster:* sterility resulting from gonadal dysgenesis in the *P-M* system. Genetics 92:1127–1140.

Laughnan, J. R. 1955. Structural and functional bases for the actions of the A alleles in maize. Am. Nat. 89:91–104.

Lawrence, Peter. 1992. *The Making of a Fly: The Genetics of Animal Design.* Oxford: Blackwell Scientific.

Lawrence, P. A., and S. M. Green. 1979. Cell lineage in the developing retina of *Drosophila.* Dev. Biol. 71:142–152.

Lawrence, P. A., and G. Morata. 1976. The compartment hypothesis. In P. A. Lawrence, ed., *Insect Development,* pp. 132–149. Oxford: Blackwell Scientific.

Lederberg, Joshua, and E. M. Lederberg. 1952. Replica plating and indirect selection of bacterial mutants. J. Bacteriol. 63:399–406.

Lederberg, Joshua, and E. L. Tatum. 1946. Novel genotypes in mixed cultures of biochemical mutants of bacteria. Cold Spring Harbor Symp. Quant. Biol. 11:113–114.

Lerner, R. A., and Alfonso Tramontano. 1988. Catalytic antibodies. Sci. Am., March:58–70.

Lewis, E. B. 1949. *su²-Hw:* suppressor-2-*Hairy-Wing.* Drosophila Info. Service 23:59–60.

Lewis, E. B. 1951. Pseudoallelism and gene evolution. Cold Spring Harbor Symp. Quant. Biol. 16:159–174.

Lewis, E. B. 1954. The theory and application of a new method of detecting chromosomal rearrangements in *Drosophila melanogaster.* Am. Nat. 88:225–239.

Lewis, E. B. 1955. Some aspects of position pseudo-allelism. Am. Nat. 89:73–89.

Lewis, E. B. 1963. Genes and developmental pathways. Am. Zool. 3:33–56.

Lewis, E. B. 1964. Genetic control and regulation of developmental pathways. In M. Locke, ed., *The Role of Chromosomes in Development,* pp. 231–252. New York: Academic Press.

Lindsley, D. L., C. W. Edington, and E. S. von Halle. 1960. Sex-linked recessive lethals in *Drosophila* whose expression is suppressed by the Y chromosome. Genetics 45:1649–1670.

Lindsley, D. L., and E. H. Grell. 1967. Genetic variations of *Drosophila melanogaster.* Carnegie Inst. Wash. Publ. 627:1–472.

Luce, Wilbur M. 1935. Temperature studies on *Bar-Infrabar.* J. Exp. Zool. 71:125–147.

Luria, S. E., and Max Delbrück. 1943. Mutations of bacteria from virus sensitivity to virus resistance. Genetics 28:491–511.

Maas, W. K. 1961. Studies on repression of arginine biosynthesis in *Escherichia coli.* Cold Spring Harbor Symp. Quant. Biol. 26:183–191.

McClintock, Barbara. 1944. The relation of homozygous deficiencies to mutations and allelic series in maize. Genetics 29:478–502.

McClintock, Barbara. 1946. Maize genetics. Carnegie Inst. Wash. Year Book 45:176–186.

McClintock, Barbara. 1948. Mutable loci in maize. Carnegie Inst. Wash. Year Book 47:155–169.

McClintock, Barbara. 1957. Genetic and cytological studies of maize. Carnegie Inst. Wash. Year Book 56:393–401.

McClintock, Barbara. 1978. Development of the maize endosperm as revealed by clones. In Stephen Subtelny and I. M. Sussex, eds., *The Clonal Basis of Development*, pp. 217–237. New York: Academic Press.

Mendel, Gregor. 1865. Experiments in plant hybridization. Verh. Naturforsch. Ver. Brunn 4 (Abhandlungen):3–47. Translated by the Royal Horticultural Society of London.

Milkman, Roger. 1962. Temperature effects on day old *Drosophila* pupae. J. Gen. Physiol. 45:777–799.

Milkman, Roger. 1967. Kinetic analysis of temperature adaptation in *Drosophila* pupae. In C. L. Prosser, ed., *Molecular Mechanisms of Temperature Adaptation*, AAAS Publ. 84. pp. 147–162. Washington, D.C.: American Association for the Advancement of Science.

Monod, Jacques, and François Jacob. 1961. General conclusions: teleonomic mechanisms in cellular metabolism, growth, and differentiation. Cold Spring Harbor Symp. Quant. Biol. 26:389–401.

Moore, John A. 1963. *Heredity and Development*. New York: Oxford University Press.

Moran, C., and A. Torkamanzehi. 1990. P elements and quantitative variation in *Drosophila*. In J. S. F. Barker, W. T. Starmer, and R. J. MacIntyre, eds., *Ecological and Evolutionary Genetics of* Drosophila, pp. 99–117. New York: Plenum.

Morgan, T. H. 1914. The failure of ether to produce mutations in *Drosophila*. Am. Nat. 48:705–711.

Morgan, T. H., C. B. Bridges, and A. H. Sturtevant. 1925. The genetics of *Drosophila*. Bibliogr. Genet. 2:1–262.

Muller, H. J. 1927. Artificial transmutation of the gene. Science 66:84–87.

Muller, H. J. 1929. The gene as the basis of life. Proc. Int. Congr. Plant Sci. 1:897–921.

Muller, H. J. 1932. Further studies on the nature and causes of gene mutations. Proc. 6th Int. Congr. Genet. 1:213–255.

Muller, H. J. 1938. The position effect as evidence of the localization of the immediate products of gene activity. Proc. 15th Int. Physiol. Congr. (Leningr.-Mosc.), pp. 587–589.

Muller, H. J. 1947. The gene (Pilgrim Trust Lecture). Proc. R. Soc. Lond. B 134:1–37.

Muller, H. J. 1950. Our load of mutations. Am. J. Hum. Genet. 2:111–176.

Muller, H. J. 1960. Remarks on the origin of life. In Sol Tax and Charles Callender, eds., Evolution after Darwin. Vol. 3: *Issues in Evolution*, pp. 69ff. Chicago: University of Chicago Press.

Noujdin, N. I. 1944. [The regularities of heterochromatin influence on mosaicism]. Zh. Obst. Biol. [Mosc.] 5:357–388. [In Russian].

Oliphant, A. R., and Kevin Struhl. 1989. An efficient method for generating proteins with altered enzymatic properties: application to β-lactamase. Proc. Natl. Acad. Sci. USA 86:9094–9098.

Owen, D. F. 1966. Polymorphism in Pleistocene land snails. Science 152:71–72.

Painter, T. S. 1931. A cytological map of the X-chromosome of *Drosophila melanogaster*. Science 73:647–648.

Painter, T. S. 1933. A new method for the study of chromosome rearrangements and the plotting of chromosome maps. Science 78:585–586.

Patterson, J. T. 1929. The production of mutations in somatic cells of *Drosophila melanogaster* by means of X-rays. J. Exp. Zool. 53:327–372.

Preuss, D., S. Y. Rhee, and R. W. Davis. 1994. Tetrad analysis possible in *Arabidopsis* with mutation of the *QUARTET (QRT)* genes. Science 264: 1458–1460.

Ptashne, Mark. 1992. *A Genetic Switch: Gene Control and Phage λ.* Cambridge, Mass.: Cell Press.

Ready, D. F., T. E. Hanson, and Seymour Benzer. 1976. Development of the *Drosophila* retina, a neurocrystalline lattice. Dev. Biol. 53:217–240.

Ripoche, J., B. Link, J. K. Yucel, K. Tokuyasu, and V. Malhotra. 1994. Location of Golgi membranes with reference to dividing nuclei in syncytial *Drosophila* embryos. Proc. Natl. Acad. Sci. USA 91:1878–1882.

Roux, Wilhelm. 1883. Über die Bedeutung der Kerntheilungsfiguren. Leipzig: Engelmann.

Rubin, G. M., and A. C. Spradling. 1982. Genetic transformation of *Drosophila* with transposable element vectors. Science 218:348–353.

Russell, L. B., and W. L. Russell. 1954. An analysis of the changing radiation response of the developing mouse embryo. J. Cellular Comp. Physiol. 43(Suppl. 1):103–149.

Sanger, Frederick, S. Nicklen, and A. R. Coulsen. 1977. DNA sequencing with chain-terminating inhibitors. Proc. Natl. Acad. Sci. USA 74:5463–5467.

Shatoury, H. H. El. 1956. Developmental interactions in the development of the imaginal muscles of *Drosophila*. J. Embryol. Exp. Morphol. 4:228–239.

Shear, C. L., and B. O. Dodge. 1927. Life histories and heterothallism of the red bread moulds of the *Monilia sitophila* group. J. Agric. Res. Wash. 34:1019–1042.

Shull, G. H. 1908. The composition of a field of maize. Rep. Am. Breeder's Assoc. 4:296–301.

Slack, J. M. W. 1991. *From Egg to Embryo: Regional Specification in Early Development.* New York: Cambridge University Press.

Smithies, Oliver. 1955. Zone electrophoresis in starch gels: group variations in the serum proteins of normal human adults. Biochem. J. 61:629–641.

Srb, A. M., and R. D. Owen. 1952. *General Genetics.* San Francisco: W. H. Freeman.

Stanley, W. M. 1935. Isolation of a crystalline protein possessing the properties of tobacco-mosaic virus. Science 81:644–645.

Stephens, S. G. 1946. The genetics of "Corky." I. The New World alleles and their possible role as an interspecific isolating mechanism. J. Genet. 47:150–161.

Stephens, S. G. 1950. The genetics of "Corky." II. Further studies on its genetic basis in relation to the general problem of interspecific isolating mechanisms. J. Genet. 50:9–20.

Stephens, S. G. 1951. "Homologous" genetic loci in *Gossypium.* Cold Spring Harbor Symp. Quant. Biol. 16:131–141.

Stern, Curt. 1960. *Principles of Human Genetics*, 2nd ed. San Francisco: W. H. Freeman.

Sturtevant, A. H. 1932. The use of mosaics in the study of the developmental effects of genes. Proc. 6th Int. Congr. Genet. 1:304–307.

Sturtevant, A. H., and G. W. Beadle. 1939. *An Introduction to Genetics*. Philadelphia: W. B. Saunders.

Sutton, Eileen. 1943. *Bar eye in Drosophila melanogaster:* a cytological analysis of some mutations and reverse mutations. Genetics 28:97–107.

Suzuki, D. T. 1970. Temperature-sensitive mutations in *Drosophila melanogaster.* Science 170:695–706.

Suzuki, D. T., A. J. F. Griffiths, and R. C. Lewontin. 1981. *An Introduction to Genetic Analysis*, 2nd ed. San Francisco: W. H. Freeman.

Suzuki, D. T., A. J. F. Griffiths, J. H. Miller, and R. C. Lewontin. 1986. *An Introduction to Genetic Analysis*, 3rd ed. San Francisco: W. H. Freeman.

Svedberg, Theodor, and K. O. Pederson. 1940. *The Ultracentrifuge*. Oxford: Clarendon Press.

Tiselius, Arne. 1937. A new apparatus for electrophoretic analysis of colloidal mixtures. Trans. Faraday Soc. 33:524–531.

Waddington, C. H. 1953. Genetic assimilation of an acquired character. Evolution 7:118–126.

Waddington, C. H. 1957. *The Strategy of the Genes*. London: Allen and Unwin.

Wallace, Bruce. 1981. *Basic Population Genetics*. New York: Columbia University Press.

Wallace, Bruce. 1983. Some apparently simple relationships involving missing scutellar bristles in *scute* mutants of *Drosophila melanogaster*. Genetica 62:147–158.

Wallace, Bruce. 1992. *The Search for the Gene*. Ithaca, N.Y.: Cornell University Press.

Wallace, Bruce, and Adrian M. Srb. 1961. *Adaptation*. Englewood Cliffs, N.J.: Prentice-Hall.

Watson, J. D., and F. H. C. Crick. 1953. A structure for desoxyribose nucleic acids. Nature (London) 171:737–738.

Watson, J. D., N. H. Hopkins, J. W. Roberts, J. A. Steitz, and A. M. Weiner. 1987. *Molecular Biology of the Gene*, 4th ed. Menlo Park, Calif.: Benjamin/Cummings.

Whiting, P. W. 1934. Mutants in *Habrobracon*. II. Genetics 19:268–291.

Wood, K. V., Y. A. Lam, H. H. Seliger, and W. D. McElroy. 1989. Complementary DNA coding click beetle luciferases can elicit bioluminescence of different colors. Science 244:700–702.

Zworykin, V. K. 1941. Image formation by electrons. Cold Spring Harbor Symp. Quant. Biol. 9:194–197.

Index

Activator-dissociation system in *Zea mays*, 205–208
Active site, enzymatic, 185
Alkaptonuria, 6–7
Amorph, 92
Anderson, Wyatt, 135
Antibodies, 185
Antimorph, 92
Arabidopsis thaliana, 16–18
Astbury, W. T., 31–33
Auerbach, Charlotte, 5
Auxochrome, 77
Avery, O. T., 41

Baker, W. T., 118
Baltimore, David, 43
Bass, B. L., 186
Bateson, William, 7
Beadle, G. W., 6, 9, 93–96, 108
Becker, H. J., 178
Beerman, Wolfgang, 153
Bell, F. O., 31–33
Benzer, Seymour, 22, 160
Berns, M. W., 34
Boveri, Ernst, 49
Brady, J. P., 222
Breakage-fusion-bridge cycle (*Zea mays*), 204
Brenner, Sydney, 22–23
Bridges, Calvin B., 3, 25, 35, 56, 58
Britten, R. J., 192
Burnett, F. MacFarlane, 215

Caenorhabditis elegans, 22, 23
Castle, W. E., 15
Catalytic antibodies, 219–222

Catcheside, D. G., 109
Cech, T. R., 186
Cepaea hostensis, 237
Cepaea nemoralis, 237
Cesium chloride gradient, 27
Chambers, Robert, 25
Chaperonin, 155
Chen, T. Y., 57
Chironomus, 153
Chromophore, 79
Compartment hypothesis, 173
Compound eye, structure of, 169–171
Conditional mutation, 145
Confocal microscopy, 34, 44
Constitutive mutant, 144
Copia-like element, 208
Corcoran, M. R., 53
Creighton, Harriet, 204
Crick, F. H. C., 1, 5, 173

Darwin, Charles, 16
Darwinian evolution, 214
Davidson, E. H., 192
Davis, B. D., 41
Delbrück, Max, 20–22, 37, 215
Delphinium, 99–100
Demerec, Milislav, 21, 110–111
Dideoxyribonucleotide, 46
Dissociator, 102
DNA library, 228
Dobzhansky, Th., 145
Dodge, B. O., 18
Dosage compensation, 91
Drosophila melanogaster, 3, 15, 35, 153, 223. *See also* Gene, mutant metabolic rate, 134

Drosophila melanogaster (*cont.*)
   temperature-sensitive lethals, 147–148
Drosophila pseudoobscura, 135, 145–147,
   223
Dulbecco, Renato, 22

Edgar, R. S., 151
Electrophoresis, 29–30
Emerson, R. A., 16
Emmerling, M. H., 99
Ephrussi, Boris, 22, 93–96, 109–112
Ephrussi-Taylor, Harriet, 41
Epistasis, 76
Epitope, 216
Epstein, R. H., 151
Escherichia coli, 20–21, 27, 40, 144
Extrachromosomal elements, 197–203
Eye tissue transplantation in *Drosophila*
   melanogaster, 93–96

Fate map, 156–166
FB element, 208
Fedoroff, Nina, 205
Fincham, J. R. S., 19
Fluorescent labels, 35
Follicle cells, 68
Formylkynurenin, 96
Franklin, Rosalind, 33

β-Galactosidase, 144
Garrod, A. E., 5–11
Gene, mutant
   Arabidopsis thaliana: *quartet*, 18
   bacteriophage: *amber*, 149–151
   corn (*Zea mays*)
      R locus (pigmentation), 97–99
      Spm (suppressor and mutator), 102
   Drosophila melanogaster
      abrupt, 92
      Antennapedia, 70, 74
      apricot, 56, 92
      aristapedia, 54
      Bar, 35, 59, 104–117, 136
      bicoid, 68, 72
      bithorax, 3, 4, 53–67
      bithoraxoid, 3
      bobbed, 56
      cardinal, 94
      cinnabar, 94, 96
      claret, 94
      Contrabithorax, 3
      crossveinless, 54, 142, 154
      Curly, 135
      cut, 54
      Double-bar, 35
      drop-dead, 163–166

      ebony, 92
      eosin, 56, 91
      forked, 92
      grandchildless, 68
      hairy, 72
      Hairy wing, 56, 93
      Hexapter, 55
      hyperkinetic, 163
      Infrabar, 136–140
      Krüppel, 72
      Microcephalus, 57
      nanos, 68
      Polycomb, 55
      postbithorax, 3
      proboscipedia, 54
      scute, 56, 92
      sepia, 94
      singed, 54
      spineless, 57
      suppressor of *Hairy wing*, 56
      Ultra Bar, 59
      Ultrabithorax, 3, 74
      vermillion, 91, 94, 96
      vestigial, 54
      white, 92
      wingless, 73
   Escherichia coli: auxotrophic mutants,
      20
   human
      albinism, 11
      alkaptonuria, 7–11
      cystinuria, 11
      hemophilia, 24
      pentosuria, 11
   mouse
      agouti (*A*), 87
      black (*B*), 87
      color (*C*), 85
      dilute (*D*), 87
      pink-eye (*P*), 87
      spotted (*S*), 87
   wheat: *Pairing homeologous* (*Ph*), 196
   yeast: *petite*, 22
Gene, unit of mutation, 34
Genetic assimilation, 135, 142
Genetic engineering, 41
Giberellin, 52–53
Goldschmidt, Richard, 5, 141–142
Gossett, W. S. ("Student"), 22
Green, S. M., 171
Grell, E. H., 136

Haemophilus influenza, 24
Haldane, J. B. S., 9, 82, 97
Hardin, Garrett, 112
Heat-shock protein, 153–155

Hershey, A. D., 14
Heterochromatin, molecular biology of, 128–130
Himalayan rabbit, 131
Homeologous chromosomes, 195
Homology, two meanings of, 193
Horowitz, N. H., 148
Hotta, Yoshiki, 160
Human Genome Project, 244–245
Hybrid dysgenesis, 208
Hybridomas, 35
Hydroxykynurenin, 96
Hypermorph, 92
Hypomorph, 56, 92

Indigofera, 80
Induction, 143–145
Insertion sequence (*IS*), 197
Intron, 43

Jacob, François, 143–145

Kacser, Henry, 1

β-Lactams, 216–219
Lamarckian evolution, 214
Lamda phage, 199–203
Landauer, Walter, 5
Laser beam, 35
Laughnan, J. R., 99
Lawrence, Peter, 70, 171, 173
Lederberg, E. M., 37
Lederberg, Joshua, 21, 25, 37, 40, 215
Lewis, E. B., 3, 62, 118
Limnaea peregra, 69
Lindsley, D. L., 127, 136
Luciferase genes as reporter genes, 213
Luria, Salvadore, 20, 21, 37, 215
Lwoff, André, 143–145

Maas, W. K., 190
MacLeod, C. M., 41
Mammalian immune system, 215
Marriages between cousins, 8
Maupertuis, Pierre, 23
McCarty, Maclyn, 41
McClintock, Barbara, 16, 99–103, 204
Melanin, 88–91
Mendel, Gregor, 14, 51
Milkman, Roger, 142–143
Mirabilis jalapa, 83
Mitotic crossing over, 166
Monilia sitophila (=*Neurospora crassa*), 18
Monoclonal antibody, 34, 220
Monod, Jacques, 143–145

Moore, J. A., 50
Moran, C., 211
Morata, G., 173
Morgan, T. H., 19, 37, 49, 56
Morphogenic furrow, 173
Mudd, Stuart, 32
Muller, H. J., 5, 7, 34, 37, 83, 91, 112, 243

Neisseria gonorrhoeae, 223
Neomorph, 92
Neurospora crassa, 6, 148
Nicotiana alata, 77
Nicotiana langsdorfii, 77
Northern blot, 44
Noujdin, N. I., 128
Nurse cells, 68

Oenothera blandina, 109
Oliphant, A. R., 216
"One-gene–one-enzyme" hypothesis, 6, 148
Operator, 144
Operon model, 144
Owen, D. F., 237
Owen, R. D., 9

Painter, T. S., 3
Palindrome, 42, 189
Papilio machaon, 140
Parasegment, 73
Patterson, J. T., 166
Pavan, Corowaldo, 153
Pederson, K. O., 28
P element, 208–211
Phenocopy, 140–143
Photosynthesis, 40
Polycistronic mRNA, and translation control, 203
Polyclone, 176
Polymerase chain reaction (PCR), 47, 234–237
Position effect, 59, 104–130
  kinetic hypothesis, 109
  structural hypothesis, 110
  variegated, 110, 117–128
Pseudoalleles, 59
Ptashne, Mark, 199
Punnett, R. C., 7
Pyrophorus plagiophthalamus, 213

Radioactive isotopes, 39
Ready, D. F., 171
Replica plating, 37
Repression, 143–145
Repressor, 144
Resource allocation, 244

Restriction enzyme, 42, 223
Reverse transcriptase, 43
Rhynchosciara, 153
Ribozyme, 186
Richmond, R. C., 222
Russell, L. B., 167
Russell, W. L., 167

Saccharomyces cerevisiae, 22
Saggitaria saggitifolia, 132
Salmonella typhimurium, 27
Sanger, Frederick, 45
Scanning electron microscope (SEM), 26
Schmalhausen, I. I., 82
Scutellar bristles, patterns of, 179–182
Shatoury, H. H. El, 57
Shear, C. L., 18
Shull, G. H., 16
Smithies, Oliver, 29, 38
Solid phase synthesis, 47
Southern blot, 44
Spassky, Boris, 145
Srb, Adrian M., 9
Stadler, L. J., 5
Stanley, Wendell, 22
Stephens, S. G., 67
Stern, Curt, 8
Streptococcus pneumoniae, 41
Struhl, Kevin, 216
Sturt (unit for fate mapping), 163
Sturtevant, A. H., 108, 156
Sucrose gradient, 27
Suppressor mutation, 56
Sutton, Eileen, 109–117

Suzuki, D. T., 147–148, 165, 243
Svedberg, Theodor, 27, 28

Target theory, 165
Tatum, E. L., 6, 21, 40
Temin, Howard, 43
Temperature-sensitive mutations, 143–153
Temperature-sensitive periods, 136–140
Tiselius, Arne, 29
Torkamanzehi, A., 211
Transgenic organisms, 222–234
Transposable elements, 183–211
Transposon, 198
Transvection, 64–67
Triticum aestivum, 193–197. *See also* Gene, mutant: wheat

Ultracentrifuge, 27

Vanessa uritica ichnusa, 140
Vanessa uritica polaris, 140

Waddington, C. H., 1, 33, 135
Wallace, Bruce, 1, 181
Watson, J. D., 2, 5, 241
Western blot, 44
Whiting, P. W., 145–146
Wood, K. V., 213
Wright, Sewall, 15, 88

X-ray diffraction, 31–34

Zea mays, 16, 97. *See also* Gene, mutant: corn
Zworykin, V. K., 26